D0698983

How "American" Is Globalization?

How "American" Is Globalization?

William H. Marling

THE JOHNS HOPKINS UNIVERSITY PRESS   BALTIMORE

The Johns Hopkins University Press
2715 North Charles Street
Baltimore, Maryland 21218-4363
www.press.jhu.edu

Library of Congress Cataloging-in-Publication Data

Marling, William H., 1951–
    How "American" is globalization? / William H. Marling.
        p.   cm.
    Includes bibliographical references and index.
    ISBN 0-8018-8353-9 (hardcover : alk. paper)
    1. Globalization—Social aspects. 2. United States—Foreign relations—2001– .
3. Civilization, Modern—American influences. I. Title.
    JZ1318.M345 2006
    337.73—dc22   2005023400

A catalog record for this book is available from the British Library.

CONTENTS

Preface

Somehow globalization is "American." Both academic Marxians and *Wall Street Journal* editorial writers tell us so, differing in their degrees of guilt or pride. This assumption, in its modest form, merely acknowledges the worldwide reach of U.S. cultural or business activity, but at the extreme it hints of conspiracy, alleging that globalization is a nefarious Americanization. Nowhere is the assumption itself questioned. This amounts to national hubris, frankly, and one result is that when globalization hiccups, critics from Paris to Tokyo feel entitled to pass the buck to the United States. And Americans shamefacedly accept it. The purpose of this book is to investigate the assumption that globalization is American. I define globalization as the increased economic integration and interdependence of nations, driven by liberalized trade and capital flows. The kinds of questions I want to ask are: Is the appetite for fast food American? Is the spread of the Internet American? Is the widespread use of English American? Just what is culturally *American* about globalization?

Such an interrogation could extend to several volumes, but some aspects of globalization are more properly the provinces of other scholars. This book does not deal with the effects wrought by the World Bank, the International Monetary Fund, or the United Nations, or by AIDS, development monies, currency devaluations, organically modified foods, and several other topics that it might properly include. Rather I have sought to focus on the culture of everyday life outside the United States and the degree to which it has been "Americanized." I try to disentangle American culture from modernity in general, for there is a large element of the latter in globalization. The speeding up of life, sanitation, hygiene, the communications revolution, urbanization—these are modern, but not necessarily American, changes to life. "Modernity" itself is examined in many other studies but usually not in relation to American culture and globalization.

So, how *American* is globalization? The short answer has two parts, and they are separated by a clutch of cultural restraints. Answer one: globalization is not as American as we think it is. It's just that everyone, especially Americans, recognizes American films, language, and logos when abroad, and draws the conclusion that the world is becoming Americanized. This is a bad case of cultural myopia. The globe is not speaking more English or dining daily at Wendy's.

In fact, a clutch of cultural constraints—language, food, habitation patterns, educational institutions, attitudes toward race and honesty—resist change in general, and resist the American face of globalization especially. They do so because change makes a bad fit with local culture. Imagine trying to park your Oldsmobile in Osaka! We should not underestimate this "persistence of the local," the ways in which local culture determines what can be globalized.

But the second and best answer to "How *American* is globalization?" seems to be "more than we know." Most of us understand little of the logistics of globalization, such as containerized freight, franchising, cash machines, bar codes, commercial aviation, and airfreight—all American-born features of modern economic exchange. They are only tools, but they suggest modes of use and, indirectly, ways to organize daily life. The use patterns attending these inventions *are* increasingly American. But even here, skepticism is appropriate. The longshoremen of Marseilles now use hand-held bar code scanners to log in containerized freight, but they still call them *marteaux,* after the hammers that their grandfathers used to remove shipping manifests. Does the use pattern or the historic allusion carry more *oomph* in daily life?

After a précis of my views, the next thing that I owe readers is a look at my passport. Gone are the days when one could write speculatively on this topic from the university library, by reading theory, or by attending conferences. So: Between 1982 and 2002 I lived outside the United States for six years. I taught at universities in Spain, France, Austria, and Japan, each for over a year. I spent six months in Mexico, and a month each in China, Poland, the Baltics, India, and Russia. For lesser periods I traveled in Holland, Belgium, Egypt, Hong Kong, Hungary, Israel, Jordan, the Czech Republic, Slovakia, Slovenia, Greece, Turkey, South Korea, Cambodia, Thailand, Indonesia, Malaysia, India, Germany, and Switzerland. I traveled alone for the most part, taking notes and pictures, riding local buses and trains, sometimes hitchhiking. I rode in the back of a pickup truck across Cambodia. I ate the local food whenever possible, even at the Hakodate fish market where it was still alive. Sometimes, yes, I ate at McDonald's. I talked to as many people as I could—I speak French, Spanish, German, and Japanese. Despite all this travel, I have focused my examples in three nations—Mexico, France, and Japan—in order to develop a thicker description of local interaction with globalization.

In a previous career I was a financial journalist for *Fortune* and then *Money* magazines, and in this book I have followed that training to find the fault lines of producers, markets, work, and consumption habits. Given this approach, I found the *New York Times, Wall Street Journal,* and *International Herald Tribune* enormously helpful, and the reader will find their reportage gratefully acknowledged. However, I also profited from the scholarship of Ronald Inglehart, Geert Hofstede, Richard Pells, John Tomlinson, Reinhold Wagnleitner, and William W. Lewis. Globalization sounds dry and abstract, but making it specific and local has been a labor of love for these writers and scholars, and I want to thank them, even as I add my two cents to the debate.

1 "Less Than We Think"

In the tiny streets of the Kitaguchi area of Nishinomiya, Japan, thousands of commuters flow in and out of the train station. You and I stand in the station entrance, hungry and looking for a place to eat. Huge neon signs cover the buildings, walls of language we don't understand. It is over-whelming and bewildering. But we do recognize the logos of McDonald's, KFC, Burger King, 7-Eleven, and Mister Donut. A bit relieved, we walk toward the Burger King, passing two boys wearing Oakland Athletics caps. It's more familiar now. There's even a girl in a Cleveland Browns T-shirt giving away free samples of tissue.

Inside the Burger King we find a menu in English and Japanese, nap-kins (a rarity here), and clean bathrooms. The counter clerk speaks a lit-tle English. This bit of familiarity, especially the fast-food logos, tempts us to conclude that Japan is being "Americanized."

What Are We Talking about When We Talk about Globalization?

Much analysis of globalization depends on these instants of recognition. What is recognized in an ocean of difference is the culturally familiar. From journalist Thomas L. Friedman to scholar Fredric Jameson, those concerned about globalization have experienced these moments. Alert to the ambiguity of signs in other contexts, they leap to form patterns when they see familiar logos or television programs or computers while abroad. Most commentary on globalization depends so much on the overdetermination of these moments that it misses how completely for-eign the foreign remains. These analyses are more revealing of a nostalgia for "authenticity" and what anthropologist Pierre Bourdieu calls the *habitus,* or customary culture, of the critic.

Consider the full realm of linguistic information outside of this Japanese train station. About two hundred signs are visible, perhaps seventy of them identifying places to eat. There are even signs in Korean and Chinese, and one in Tagalog. In the three scripts of Japanese there are signs for sushi shops, noodle shops, bars, dumpling stands, *okonomiaki* and *yakitori* restaurants, and *ramen* counters in a nearby department store and in the train station. There are canteens for railroad employees and taxi drivers, even restaurants *without signs,* identifiable by smell, word of mouth, or the cooking utensils left out to dry. Within four hundred yards, of the more than seventy places to eat or drink, we recognize the logos of only five.

In the Burger King, the only other customers are a mother and daughter at a table and a businessman waiting for takeout. The delivery boy packs two orders on his motorcycle. Business is not exactly booming. Looking at the menu, we see burgers with *teriyaki, wasabi,* or *shichimi* flavorings—pretty innovative for an American company, right? Actually, Burger King, though it began in the United States, is not "American"; it is owned by Diageo of Great Britain. Nor is there much American about it in the culinary sense, because some form of fried beef exists in most cultures—witness the *shabu-shabu* franchise around the corner, and down the block a Moos Burger, a Japanese hamburger chain.

The other "American" logos in view are similarly slippery. The logo of 7-Eleven connotes familiarity, but it is now a Japanese company. The Ito-Yokado Corporation saved 7-Eleven from bankruptcy in the 1970s and now it owns them all, even the one in your neighborhood. Down the street is 7-Eleven's main rival, Lawson's, which has a familiar green and blue logo that may look American but is also Japanese. Likewise the Mister Donut. It sells some donuts, which cost one to two dollars each, but many more breakfasts of *miso* and rice porridge. No cops on coffee breaks there: Mister Donut's customers are "office ladies" in search of a quick, genteel breakfast or lunch. Its promotional giveaways are umbrellas and tote bags.

At least the McDonald's is American, right? No, not really. Ever since it began to sell burgers on Tokyo's Ginza in 1971, McDonald's has been a joint venture with legendary Japanese businessman Den Fujita. In 2001 McDonald's spun off the Japanese subsidiary to Fujita and other investors for about $1 billion. All the McDonald's in Japan are now Japanese owned.

Not that this would surprise millions of Japanese children and teenagers, who have always thought of it as a Japanese company.

What do logos reveal about globalization? Not much. They connote familiarity, make us believe that we will find a consistency in the products sold under them. If they fulfill their "selling promise," logos can be worth millions, sometimes more than the physical assets of the companies that own them. But other logos vanish overnight—remember Kresge and Sohio? Nor were logos invented in the United States; they've been around at least since Julius Caesar's face identified Roman coins.

Rather than instant identity, foreign logos can be the source of cultural confusion. When Merrill Lynch opened neighborhood offices in Japan, its red bull logo persuaded many that it was a Korean barbecue shop. It lost $600 million and closed the offices three years later. The California software firm Niku, which generates 40 percent of its sales abroad, could not understand why its products weren't selling in Japan. Someone pointed out that *niku* means "flesh" or "meat" in Japanese. Kind of creepy, a software program called "flesh." But *niku* means "one who does good" in Farsi, protested company founder Farzad Dibachi. He has since changed his products' names: "We need a name that isn't a food group in an important geography." At the same time, Japanese logos confuse foreigners: Would we buy a sports drink named "Pocari Sweat," a coffee creamer called "Creap," chocolates "Colon" and "Negro," or the soft drink "Calpis"? The arbitrariness of logos was illustrated by a *New York Times* reporter in Iraq who found a restaurant called "Italiano," which served no Italian food but did have a dish called "Kentucky." Neither the owner nor patrons could say what "Kentucky" meant, but they swore by the dish.

Logos are less meaningful than we suppose. True, companies spend millions creating them, and logos do help McDonald's and Nike to sell goods, but there are large companies, like Procter & Gamble, whose logos we cannot easily conjure up. A logo works only if we connect it to a "use-habit" in daily life, where it becomes "naturalized" as part of our environment. And this competition for our use-habit attention (and our money) is basically local and rooted in cultural specificity. When we enter a British Petroleum or Royal Dutch Shell gas station in the United States, we are not Anglified or Dutchified. We are buying convenient, dependable gas, and maybe a car wash or quart of milk. Price, reliability, and service top our criteria. Every successful logo, domestic or foreign, has hacked out a long and arduous path to such "reception" by users. Think of the trail Toyota trod in the United States to raise the public estimate of its cars from "cheap" to "quality." A foreign logo in the United States, like Au Bon Pain in a Manhattan subway station, gets us to supply an appropriate context for it only after we have made a place for French bread in our lives. In arguments about globalization, however, U.S. logos are taken as evidence of cultural imperialism. We all *see* Visa, McDonald's, KFC, Coke, and Wal-Mart. Mickey Mouse grins at us in France and Japan. Bruce Willis frowns from billboards and posters. *CSI: Miami* and *Everybody Loves Raymond* appear on the hotel's cable channel. The taxi drivers and hotel employees speak a little English. All this seems part of a creeping "Americanization."

THE MANUFACTURE OF LOGOS

Our semiconscious supposition is that the kids we passed wearing Oakland Athletics and Cleveland Browns gear understand, if only partially, the logos: the boys know the Athletics play baseball, the girl that the Browns play football. But when we stop and ask them, we find that they don't. The girl doesn't know what Cleveland is, period. I've been asking logo wearers about their apparel for seven years, in dozens of countries. People everywhere admit to wearing logos for their cachet, for their novelty, just as New Yorkers in 1999 wore baseball caps with Japanese characters on them, just as U.S. basketball players Allan Iverson and Marcus Camby tattooed their bodies with *kanji*.

Logos, T-shirts, and tattoos are reminders that many of the signs that circulate among us are empty. In France in 2002, many trendy shirts were emblazoned with Arabic script. When I asked about the meaning, the wearers were all clueless. In the "Arab market" of Avignon, I asked a Frenchman selling these shirts to translate one. He confessed that he didn't know *exactement*, but he bought his stock from an Italian in Nice, who got them from a Lebanese supplier, who manufactured them in Turkey. "It is *his* name," said my informant of his supplier, "like Richard or Perreau."

This anecdote opens a window on the production of "logowear," which will take us beyond the "cachet" of the street. Sewn in Turkey or Pakistan or Vietnam or Mexico, logowear now appears on the streets of those countries as frequently as in the developed world. Why? Because production contracts shift between small factories, sometimes monthly, leaving managers with excess stocks and little allegiance. Seconds and blemished goods cannot be shipped. Pilferage is a fact of life, and without job-site control by the logo owner, knockoffs are possible and practical, especially using cheaper materials. Enforcement of copyright laws is practically nonexistent. The police arrive if Gucci handbags or Hermes scarves are counterfeited, but only when those firms make a stink.

A STROLL DOWN PUENTE DE ALVARADO

Mexico City is a typical birthplace of and graveyard for logowear. On Puente de Alvarado in 1997, street vendors were selling knockoffs of NBA and English soccer jerseys, Nike and Adidas caps, Levi's and Jordache jeans, not to mention watches, shoes, and handbags. The prices were rock bottom. Though Mexicans wore fewer logos than Americans, there were "American logos" all around Mexico City, many taken incongruously out of their American status contexts. In one hour, I saw a grown man wearing a Bugs Bunny T-shirt, a teenager in the sweltering subway wearing a Penguins hockey T-shirt, and an emaciated, five-foot-tall gum vendor wearing a Shaquille O'Neil T-shirt. Another young man sported a DEVO T-shirt, Chicago Bulls sweatpants, and a Raiders football cap.

Some of this logowear was the residue of *maquiladoras,* clothing plants on the U.S. border. These were irregulars, odd lots, and discontinued styles: the old Oakland Raiders colors or Michael Jordan as number 43 or the 1994 World Cup. On Puente de Alvarado the 1997 prices ranged from four to eight dollars for a pair of "American" logo denim jeans, to two to five dollars for a logo T-shirt. For most buyers, these were simply garments; if they could have afforded more, they would have bought new T-shirts touting Mexican singing sensation Ana Barbera or soccer star Hugo Sanchez.

Rather than being driven by "Americanization," however, such logowear exists in the context of the Mexican clothing status system, which begins with shoes—do you have any? If you do, you're already one step up the status ladder, because many Mexicans do not. Barefoot *indigenos* ride the subway and walk the streets. At the top of the ladder are middle- and upper-class university students in the San Angel neighborhood of the enormous Autonomous National University of Mexico (U.N.A.M.). In 1997 a version of Euro-Yuppie attire was ascendant there, depending heavily on sweaters, sports coats, ties, pleated pants, and leather shoes. It was all deliberately logo free—not difficult to understand when the *maquiladoras* are churning it out. Argyles, plaids, and patterns were popular. From U.N.A.M. and the eateries and copy shops around the Copilco Metro, such styles may then spread through San Angel to Coyoacan. If they appeared in the McDonald's at Centro Coyoacan, or at the McDonald's in the Zona Rosa, then a trend was afoot. But the chances of argyle making it that far were slim.

The Mexico City fashion system keeps foreign logowear at the mid to lower levels. On Puente de Alvarado we could find "Jurassic Park" shorts or "Deutschland" World Cup T-shirts. The vendors sold these alongside used American clothes—*ropa usada*—smuggled into Mexico for resale. It has to be smuggled because importing used clothing to Mexico is illegal; it takes jobs away from Mexicans. This is contraband from the North.

This trade is worth a footnote, because it illustrates just how different something as basic as a T-shirt can be from culture to culture. The *ropa usada* trade centers along the Texas-Mexico border (the *maquiladora* zone), particularly Brownsville, where nearly two dozen U.S. entrepreneurs specialize in it. The King of Used Clothes is sixty-eight-year-old Jim Johnson, who oversees the purchase, sorting, and resale of tons of clothing rejected by Goodwill Industries, the Salvation Army, and St. Vincent de Paul. Johnson buys tractor-trailer-loads of old shirts, pants, and dresses for pennies on the pound. "I call it recycling," says Johnson, winking at Mexican law.[1] He bought his first *ropa usada* store in 1964 for $400. He understands the local context of clothing: he sends cotton outfits to Kenya and Nigeria, vintage clothes to Japan and Italy, polyester to India and Pakistan, rags to oil field crews. His operation fills a 170,000-square-foot warehouse, employs 350 people, and processes more than fifty tons of used clothes a day. "You'd think we'd run out of clothes," he says. "But we never do."

The going rate for clothes not selected for the specialized markets was around $1.25 a pound in 1996. A typical buyer plunked down a $300 deposit for the right to open one of the fresh bales, each of which sold unopened for $1,000. The buyer can pick out the treasures, up to 250 pounds, before rummagers called *chiveros* attack the carcass of the bale. To carry the goods to Mexico, the buyer and *chiveros* each pay $25 to a *pasador*—a runner—who will carry the goods over the international bridge. Then the buyer will bribe his or her way through four other government checkpoints, paying about $150 before getting to Puente de Alvarado. Ironically, a portion of what the smugglers bring back was made in Mexico. Mexican labor was paid to make these items, which were exported to the United States and sold at retail. Then the items come back to Mexico used and are resold at a fraction of original cost. Clothing would cost more in Mexico, obviously, if there were neither *maquiladoras* nor clothing smugglers. The antismuggling laws were designed to get Mexicans to buy retail and to protect *maquiladora* jobs.

Logowear is also common on Puente de Alvarado because of counterfeiting. Ever since Levi Strauss lost the "Great Jeans Battle" around 1990,

Mexico has become adept at cloning clothing. Levi's was a patriotically "American" company, with eleven factories in four states and 31 percent of the American market in denim. Then Gap, Guess, Calvin Klein, VF Corporation (maker of Lee and Wrangler), and even J. C. Penney and Sears (with Arizona and Canyon River Blues, respectively) created an annual model change in jeans that the button-fly patriarch disdained. Foreign companies such as Diesel (Italy) chipped away at Levi's upscale market with $99 jeans. Many of the rival jeans were made in Mexican factories, so it's hardly surprising that the craft of jeans making became widespread. Counterfeit clothes moved north more easily than drugs. The trained eye can recognize them in stores in Los Angeles, Houston, Dallas, Denver, or Chicago, but the differences may be minute, as inconspicuous as the stitching patterns inside pockets. Even buyers for major chains have been fooled about the provenance of $200 jeans. Logos, who can keep up with them? Is "Diesel" an Italian, Mexican, or American logo? Does the logo *mean* anything?

Critics think so. One writes that "this form of globalization is actually Americanization, because the United States is by far the biggest producer of pop culture goods."[2] But does the fact that logos are in English (even if owned by Italians) really mean that the circulation of logowear is Americanization? Pop culture goods supersede one another rapidly and signify different things in different cultures. When they attain any meaning at all, they are simply placeholders for novelty, status, or consumption, which are all *local* significations arising from local circumstances.

When I returned to the train station at Kitaguchi, Japan, in 2003, I found the constellation of logos was the same size, except that Wendy's and Mister Donut were gone. KFC and McDonald's had survived, but a huge new department store and apartment building overshadowed the neighborhood. Its ground floor was a warren of eating places, and not a single logo was American.

Is English Conquering the World?

Outside the Nishinomiya train station, there was a big sign for the NOVA English school, and down the block a Smith's English school was scheduled to open. English is officially the second language of Japan. Scholarship and research everywhere are published in English, and the Internet seems to be all English. Is English dominating the globe? Many critics claim so. As Burton Bollac writes, "Not even Latin, the European scholarly language for almost two millennia, or Greek in the ancient world

before it, had the same reach. For the first time, a new language, English—
a bastard mixture of old French dialects and the tongues of several Ger-
manic tribes living in what is now England—is becoming the lingua franca
of business, popular culture, and higher education across the globe."[3]

As in the case of logos, there is less here than meets the eye. Yes, English is becoming the lingua franca of business and research, for reasons
that will become clear. But English is not becoming the language of pop
culture or daily life outside traditionally English-speaking areas. In fact,
the percentage of the world population that speaks English as a native
tongue is declining. Even the percentage of Americans who speak English as a native language is declining.

Critics of globalization tell us differently. When the World Conference
on Linguistic Rights met in Barcelona in 1996, it concluded that, like
McDonald's and Starbucks, English was "colonizing" the world. Most of
the attendees were, not surprisingly, native speakers of English, French,
or Spanish who study or lobby for "endangered languages"—people with
professional interests in, say, Maori, Taino, or Apache. Like Duke University's Fredric Jameson, they believe that languages are "all equal" and
should "freely produce [their] own culture according to [their needs]."[4]
What was needed, as Romance-language scholar Walter Mignolo put it,
was action against the "hegemonic power of colonial languages" such as
English.[5]

Mignolo argued that languages from Apache to Wolof, even Ukrain-
ian and Portuguese, are being "marginalized" and may soon become
"extinct." This argument may be familiar to readers of *Time* and *National
Geographic,* for the word "extinct" implies an analogy to Darwinian evolution, and we are supposed to infer that a great number of languages,
like great diversity of fauna and flora, is healthy. But paleolinguists tell
us that only eight thousand years ago all of our ancestors spoke one
murky language called proto-Indo-European. Languages began to diver-
sify around 6000 BCE (before the common era) following the great
human migrations. Some languages soon died out, and whole language
families, such as Hittite in Turkey and Tokharian in Central Asia, disappeared four thousand years ago. But languages are constantly appearing
and disappearing. Only six hundred years after Chaucer wrote, no one
speaks Old English, but millions of people speak Bahasa Indonesian,
which did not exist in Chaucer's day. As Madelaine Drohan and Alan
Freeman point out, smaller languages have been overwhelmed by Mandarin, Spanish, Arabic, and Russian as much as by English. The best

chance for saving some of them, or prolonging their lives, turns out to be U.S.-generated computer software.

ENGLISH AS LINGUA FRANCA

It is easy to forget that Latin, Spanish, and French were true colonial languages long before the spread of English. Rather than being adopted by trading partners voluntarily, these languages were often imposed by brute force, by slave owners, Jesuits, and centurions. Nevertheless they proved useful in uniting and modernizing vast impoverished areas. But two thousand years later Latin has disappeared as a daily language, and Greek as a colonial one. Spanish remains the language of some former colonies, but has almost disappeared from the Philippines. French is declining throughout former French colonies.

As Drohan and Freeman and Joshua A. Fishman point out, English has achieved dominance only where Anglophone settlers were the majority. Despite a century of colonization, India returned to indigenous languages within fifty years of independence, even though as influential a leader as Nehru argued that English should be continued, to enfranchise all groups through a common tongue. Samuel Huntington noted in *The Clash of Civilizations* that in India "there were 18 million English speakers in 1983 out of a population of 733 million and 20 million in 1991 out of a population of 867 million . . . relatively stable at about 2–4 percent."[6] It is the various forms of Hindi that unite India today, a telling demonstration of the failure of English as a "colonial" language.

In the rest of the former British Empire, the results are similar: in Pakistan, English has been replaced by Urdu; in central Africa, by Swahili; in Hong Kong, by Cantonese. This is true of other "colonial" languages. In North Africa, Arabic has largely replaced French. In the former Soviet Union, a dozen languages have replaced Russian. "The percentage of the world's population speaking the five major Western languages (English, French, German, Portuguese, Spanish) declined from 24.1 percent in 1958 to 20.8 percent in 1992," notes Huntington.[7]

A statistical overview of languages contains other surprises. More than 40 percent of the world is still illiterate, and most of the illiterate population lives in areas where multiple, small languages are spoken. That these are generally the world's most impoverished areas should give pause to those who employ the evolution analogy. The ideals of increasing literacy and decreasing poverty seem to be incompatible with preserving small languages, especially when they have no written forms.

TABLE 1

Total Number of Native- and Second-Language Speakers
(in millions), 1980 and 1990

Language	Total speakers in		Rate of Growth
	1980	1990	
Mandarin	690	844	+22
English	380	437	+15
Russian	259	291	+12
Spanish	238	331	+39
Hindi	230	338	+47
Arabic	142	192	+35
Portuguese	141	171	+21
Bengali	140	181	+29
German	120	118	−9
Japanese	115	124	+8
Malay-Indonesian	106	138	+30
French	100	119	+19

SOURCE: *World Almanac.*

That the dominant languages of the world are shifting we can see by tracking the numbers of people speaking them. Table 1 begins with the languages ordered by total number of native- and second-language speakers in 1980 and shows that by 1990 three languages—Hindi, Arabic, and Spanish—were growing much faster than other languages. Table 2 re-orders the languages' rank for 1990 with the changes to 1999, the latest year for which information was available, and the rate of growth. Counting speakers of languages is tricky because of dialects. In China alone in 1999, there were about 220 million speakers of Cantonese, Wu, Hakka, and Min, among other dialects. To account for 75 percent of the world's population, we would have to raise the number of languages to twenty-five. To count 95 percent of world population would require one hundred languages. In other words, 20 percent of the world population speaks seventy-five smaller languages. And as Fishman points out, "never before in history have there been as many standardized languages as there are today: roughly 1,200. Many smaller languages, even those with far fewer than one million speakers, have benefited from state-sponsored or voluntary preservation movements."[8]

Now let's look at the right-most column in table 2, which I labeled "Native Speakers." The languages with the most native speakers were:

TABLE 2

Total Number of Native- and Second-Language Speakers (in millions), 1990 and 1999

Language	Total Speakers in 1990	Total Speakers in 1999	Rate of Growth	Native Speakers
Mandarin	844	1,075	+27	885
English	437	514	+17	347
Hindi	338	496	+47	375
Spanish	331	425	+28	358
Russian	291	275	−6	165
Arabic	192	256	+33	211
Bengali	181	215	+19	210
Portuguese	171	194	+13	178
Malay-Indonesian	138	176	+28	58
Japanese	124	126	+3	125
French	119	129	+8	77
German	118	128	+8	100

SOURCE: *World Almanac.*

1. Mandarin, 885 million
2. Hindi, 375 million
3. Spanish, 358 million
4. English, 347 million

English had the slowest growth rate of these languages. Spanish, Hindi, Arabic, and Bengali are all growing far faster, and if we simply extrapolated from their growth rates, they could be expected to surpass English around 2040. But the populations of Bengali and Hindi speakers are so constrained by geographic limits and resources that this is not likely. Arabic and Spanish speakers, on the other hand, face no such limits. But "Arabic speakers" is misleading, because while the speakers share a written language, spoken dialects of Arabic are often mutually incomprehensible.

By comparing the "1999" and "Native Speakers" columns in table 2, however, we can get a sense of which languages function extensively outside their home ports. And we can also see that English is not the world's most widely spoken second language.

1. Mandarin, 190 million
2. English, 166 million

3. Hindi, 121 million
4. Malay-Indonesian, 118 million
5. Russian, 110 million
6. Spanish, 67 million
7. French, 52 million

Mandarin is required in government, business, or education in China, where 200 million people speak some other language. But it is also a lingua franca throughout Asia, with strong speaker enclaves in Singapore, Indonesia, Malaysia, Japan, and even India. Hindi is the lingua franca of India. Malay-Indonesian unites speakers of two hundred local languages, some substantial in themselves (Javanese has 64 million speakers), who live on an archipelago of seven thousand islands. Russian is still spoken widely through the former Soviet Union and its satellites.

We need to cross-reference these figures with the number of nations in which a language is spoken by native- and second-language users. Let's revisit in table 3 the list of native speakers, adding second-language speakers, and the number of countries the language is spoken in, for the year 2000. Seen in this light, English is something less than a juggernaut. As far as its reach goes, it is the first language of at least some speakers in 104 countries, followed by French, spoken in 53 countries, and Spanish, spoken in 43. This makes these significant diasporic languages. In contrast Bengali was spoken in only 9 countries, Hindi in 17 nations.

TABLE 3

Total Number of Native- and Second-Language Speakers (in millions), and Number of Countries in Which Each Language Is Spoken, 2000

Language	Native	Second	Total	Countries
Mandarin	885	190	1,075	16
Hindi	375	121	496	17
Spanish	358	67	425	43
English	347	166	513	104
Russian	165	110	275	30
Arabic	211	45	256	ca. 30–40
Bengali	210	5	215	9
Portuguese	178	16	194	33
Japanese	125	1	126	26
French	77	52	129	53
Malay-Indonesian	58	118	176	ca. 17–30

SOURCE: *World Almanac.*

But *which* countries a language is spoken in is important. Mandarin, the world's leading language, is spoken in two dozen countries, including the United States. There are far more first-language Chinese speakers (1.6 million in 2000) in the United States than there are English speakers in China. More than 167 million people in 30 countries speak Russian. And just off this list are 80 million Korean speakers who live in 31 mostly developed nations. More than 60 million people in 35 countries speak Turkish, many in Germany. The presence of Japanese in 26 countries is not due to its minuscule 1 million second-language speakers, but to Japanese posted overseas by their companies. These languages and those in table 3 will be important in the future, as the world moves toward what Fishman has called regional languages: "Popular writers, itinerant merchants, bazaar marketers, literacy advocates, relief workers, film makers, and missionaries all tend to bank on regional lingua franca whenever there is an opportunity to reach larger, even if less affluent, populations. In many developing areas, regional languages facilitate agricultural, industrial, and commercial expansion across local cultural and governmental boundaries. They also foster literary and formal adult or even elementary education in highly multilingual areas."[9]

The trend of these statistics is in line with what Huntington showed using statistics from 1958 to 1992. Extending the analysis to 1999, we see that the percentage of the world's population who are *native speakers* of English actually declined from 9.8 to 7.6 percent. The percentage of native speakers of the world's leading language, Mandarin, also declined slightly, from 15.6 to 15.2 percent. But there are still twice as many Mandarin speakers as English speakers, and in real terms the number of Mandarin speakers more than doubled, from 444 million in 1958 to 907 million in 1992 (or 1,075 million in 1999). The language groups that have increased dramatically as a *percentage* of the world population are Arabic and Bengali, which each accounted for 2.7 percent of the world's speakers in 1958, but rose to 3.5 percent and 3.2 percent, respectively, in 1992. Hindi speakers rose from 5.2 to 6.4 percent, and Spanish speakers from 5.0 to 6.1 percent. English as a first language has fallen from its mid-century position of second place to fourth as the millennium ended.

Even in its most populous port, the United States, the percentage of the population described as native speakers of English has declined. According to initial reports of the 2000 U.S. census, 17.3 million Americans identified themselves as *primarily speakers of Spanish,* and that was probably underreported. Hispanic-Americans were 12.8 percent of the population in 2003, up from 9 percent in 1990. Initial analysis showed an additional 4.5 million residents whose primary language was Asian or Pacific, 3 million Arabic speakers, 2 million American Indian language speakers, 1.8 million Korean speakers, 1.5 million Czech speakers, 1.4 million speakers of Philippine languages such as Tagalog, 900,000 Italian speakers, and 800,000 Japanese speakers. Other English-speaking nations, especially Great Britain, also have large populations for whom English is a second language. This is one reason the list of native-language speakers shows English in fourth place.

In 2004 the U.S. Census Bureau revised these figures upward. The number of foreign-born U.S. residents and their children rose to 56 million in 2002, as 20 percent of the U.S. population spoke a language other than English at home. More than 28 million Americans use Spanish as their primary language (though half of them reported speaking English "very well"). By 2004 the largest group of Internet-using Spanish speakers in the world resided in the United States, which is why Yahoo, MSN, eBay, and Google introduced Spanish versions. There were parts of California, New Mexico, Texas, Florida, and New York City where Spanish was the main language. Not surprisingly, this globalization of the United States has been accompanied by a decline in English-language newspaper readership, which had long been a culturally unifying force. From 1993 to 1997 alone, U.S. English readership fell 5.2 percent. By 2000 there were 50 percent more newspaper readers in Japan than in the United States. For a "colonial" language, English seemed to have a weak and fading grip on its main base.

The claim that English is a "colonial" language seems far-fetched, especially in light of these trends. The real subject of the argument is the world's lingua franca, which Huntington defined as any language widely used as a means of communication among speakers of other languages. The original lingua franca was a polyglot of Italian and Provençal, with bits of Spanish, French, Greek, Arabic, and Turkish, and was spoken by merchants and traders to facilitate commerce in eastern Mediterranean ports in the 1600s and 1700s. Where it is spoken today outside Anglophone countries, English tends to be the lingua franca of business, education, science, and diplomacy. But it is not spoken on the dock, like the original lingua franca. What bothers academics is the use of English in their specialties. Mignolo intimates that valuable contributions might come from authors writing in the Quechua, Mayan, or Seri Indian languages. But, as Fishman notes, the argument for "languages such as Breton serves a strong symbolic function as a clear marker of 'authenticity.'"[10] Somehow, professors imagine, use of these languages is more authentic. Using English as a lingua franca does not allow them *le mot juste*, even though it allows communication across cultures.

But communication is the point of what we might call *Engla franca*. And its limit. When an Indian factory owner and a Taiwanese financier negotiate a contract in English, or when German engineers and Spanish technicians discuss an Airbus design in English, they use an *Engla franca*, but their discussions hardly carry the weight of Anglo-American civilization. They are not speaking from a colonized mind-set. As their lingua franca, English is simply a tool, like the spreadsheets they use, like the equations they consult.

Native speakers everywhere are impressed by the ostensible difficulties of their languages. English is not simple, but it is not that difficult to learn. Its teaching has been very effectively systematized in ESL (English as a Second Language) and TESOL (Teachers of English to Speakers of Other Languages) programs, as well as a British system, that are in place worldwide. In fact, only French has such an extensive and tested pedagogy for nonnative speakers. English also has the advantage of business familiarity: the trading civilization of Great Britain made English terms of business, such as pounds, tons, tare weight, and FOB (free on board), into the common parlance of shipping, negotiation, and contracts. Popular music, film, and television have given some words a similar currency, though less and less as these media become more and more local. Just as the original lingua franca was not the language that merchants of Marseilles and Beirut spoke in the bar or at home to their families, however, the speakers in my examples don't speak *Engla franca* outside of work.

This partial, regional use, and its utilitarian end have fostered local English dialects. Thus the Englishes spoken in India, Nigeria, or Singapore are often mutually incomprehensible. In the latter, efforts of the authoritarian government to diminish "Singlish" in favor of pure English have failed, and Singlish is being promoted by radio, television, and rap music.

Even in international communication, a kind of pidgin *Engla franca* must be used. This is a grammatically simplified, idiom-and-metaphor-free subset of ESL English, which Americans are sometimes at a loss to understand. Max Watson, director of an MBA program at Ohio's Baldwin-Wallace College, witnessed an American Airlines flight attendant who could not understand an Indonesian passenger speaking this idiom, until a nearby Japanese woman rephrased the requests in American English. "English is the international language of business," says Watson, but "it is no longer the U.S. version that everybody strives to speak."[11] The English that Americans speak is too idiomatic and contextual; "right" has several meanings in American, but in ESL it is only a direction. "English that flows off the tongue throws non-native speakers for a loop," says Gary Yingling, director of Asian Operations at Rockwell Automation, using two idioms he avoids overseas.[12] ESL *Engla franca* simplifies tenses, drops adjectives, and omits a/an distinctions. "Time is money in the business world," says Chris Folino, a Hong Kong business consultant: "Using only English nouns and verbs is often the clearest way to be understood."[13]

Another critique of English frets over its words entering foreign languages. The French are vigilant that "French fries" not replace *pommes frites* and have invented *logiciel* to replace "computer chip." In Brazil, some politicians wanted to authorize their Academy of Letters to legislate like the *Academie Française,* but they were taken aback to learn that many of the 400,000 official words of Brazilian Portuguese, such as *azeite* (olive oil) and *futebol* (soccer) were imported (Arabic, British English). They were most concerned about words like "upgrade" and "happy hour," for which there were no Portuguese equivalents. Government language watchdogs have not worked very well anyway.

Countries without government guardians of linguistic purity often seem to critics to be wide open to linguistic subjugation, but they forget that there is usually a counterbalancing local linguistic chauvinism. The Japanese consider their language to be so difficult that no foreigner can ever speak it correctly. When sound film appeared in Japan, the populace did not trust the semblance of foreigners speaking Japanese (dubbing), so highly paid *benshi* stood in front and explained the action, just as they had during the silent-film era. The *benshi* also broke up theaters and beat uncooperative theater owners—local linguistic chauvinism takes many forms.

LANGUAGE MISCEGENATION

Today the Japanese use English imports such as "boyfriend" (*boyfrendou*), "taxi" (*takushi*), and "door" (*doa*). They also borrow English for problems they cannot quite own up to: *homuresu hito* (homeless people), *reipu* (rape), *doraggu* (drugs), *hakku* (computer hacking), and *sekuhara* (sexual harassment). But English is hardly taking over Japan, despite the impression of visitors. Many words that visitors think they hear are false friends, such as *konsento* (a light socket) and *jokki* (a mug). Historically Japanese has drawn on many cultures, especially Chinese, but also Portuguese (*pan* is Japanese for bread, *braja* for bra), Dutch (*retteru* is the word for letter), and German (*arbeito* is a part-time job). But these imports are the slightest of surface ornamentation on deeply nativist language practices.

English speakers don't seem to realize that their language also is a mix of imported words, including Japanese. Americans use *tycoon, typhoon, honcho, karaoke, kamikaze, tsunami,* and *sushi* as if they were native words. *Hentai* and *anime* are in the vocabularies of most college students. The fifty thousand U.S. Marines stationed in Okinawa have been bringing back Japanese words for more than fifty years. The sources of American English go far beyond Japanese, however. The language includes *mattress* and *algebra* (Arabic); *yogurt, tulip,* and *jackal* (Turkish); *bazaar* and *caravan* (Persian); *shampoo, dungarees,* and *pajamas* (Hindi); *lemming* and *ski* (Norwegian); *hamburger, snorkel,* and *waltz* (German); *kayak* and *igloo* (Eskimo); *mosquito, siesta,* and *lasso* (Spanish); *ballerina, soprano,* and *casino* (Italian); and *cuisine, boutique,* and *chauffeur* (French). Americans have "naturalized" these words by hearing and then using them. There are many nations whose language cultures are just as heterogeneous, and some (Malaysia, Indonesia, the Netherlands) that have borrowed even more widely.

The limited impact of *Engla franca* is evident if we return to Nishinomiya, venture down the street, and look into that NOVA English School. The first thing a visitor notes is that students entering and leaving are not speaking English. Inside in classrooms we hear rote lessons in lowest-common-denominator English: "Where is ——?" "How much ——?" and "When is ——?" As Will Ferguson pointed out, this instruction leads only to the English equivalents of Japanese cultural positivism: "Let's all live happy life together." He cites a Japanese pharmaceutical company whose advertising department, evidently graduates of such a school, came up with the ad slogan "Let's All Enjoy Tampon Life!"

Schools like NOVA are typical of overseas commercial English instruction. They have four kinds of students. First, there are Japanese who want to avoid being fleeced on vacation: they learn "How much is the room?" "Is there a bus to Disneyland?" "Where is the Japanese restaurant?" The second kind of student has to deal with foreigners in Japan: "The station is two blocks ahead." "Do you know Japanese bathing custom?" "Can you eat *natto*?" The third kind studies English as a hobby (the Japanese are formidable hobbyists). Finally, the largest group is students, credentialing themselves as they slog down the path to a job. The instructors are young, underpaid, mostly amateur teachers from the United States, Australia, and Britain, who concentrate on Japanese interests, from Doraemon to David Beckham. Few of the students will ever use their English outside the classroom, because Japan—to relieve citizens of stress and to get foreigners into the flow—has signposted in English most places that *gaijin* want to go. To state the obvious, students at NOVA do not read the Declaration of Independence or Lincoln's Gettysburg Address in English classes. Neither do students in university courses. When I taught at a respected Japanese university, my graduate students in American literature had not read *The Great Gatsby*. But they told me it was a "very Japanese book," because a Japanese writer, Haruki Murakami, referred to it often in his novels. No Japanese on the street could understand President George W. Bush speaking. In sum, the English taught abroad is not only free of the American and British value systems but commonly skewed to local interests.

The Ubiquitous American Film

On Academy Awards night, millions of people around the world watch to see which films and actors will win Oscars. Some will carry away from the gala an impression that film is "American"—or so the producers hope. But the venerable Oscar has a big crack. As the *New York Times* detailed in 2004, Hollywood films aren't "American" anymore: they have international casts, are filmed abroad, and don't reflect American values. "People used to go to American movies and want to be American," wrote editor Lynn Hirschberg; "Now you go to a movie and you can't even tell what an American is."[14]

To understand what has changed, we need to look at markets. The world's second largest film market is Japan, so let me return to the commuter train station in Nishinomiya again. Even though 100,000 people pass daily through here, there are no theaters in this upscale neighborhood. Japan's cultures of film, from production to attendance, are local and very different.

In the newspaper, we can find half a dozen "Hollywood films" playing in Kobe or Osaka, each thirty minutes away by train, but an equal number of Japanese films, and a few Chinese. There were only 2,585 screens in Japan in 2002, most owned by the Toho and Shochiku chains. People attended a film only 1.3 times a year, the lowest rate in the industrialized world. By contrast, Americans attended an average of 5.3 films per capita that year. So how can this be the world's second richest market? The answer is that tickets cost eighteen dollars and up, and the only bargains are Tuesday matinees for women at fifteen dollars. And video rental stores—there are two outside the station—do a phenomenal business, with hundreds of Hollywood titles for rent. Rentals are over 70 percent of the market in Japan.

Let's begin with the obvious—film is one of the leading U.S. exports. Just as France exports wine and Thailand rice, the United States produces a high-quality, inexpensive commodity. No one gets excited about the amount of Australian wine consumed or number of Hong Kong kung fu films shown in the United States. Despite its highbrow pretensions, film is basically entertainment. But critics of globalization exclaim that film is much more than that: it is a pillar of local language and culture, almost essential to self-expression. Even though the biases of academic critics seem to speak louder in this protest than that of the "subaltern" audience, let's explore this argument. What is actually playing on screens in Osaka? Do the films of Steven Spielberg stifle the cultural expression of the Japanese? Is there a local film industry? Is it being thwarted by "Hollywood"? As we will see, "Hollywood" is internationally owned and operated, spending money and returning profits all over the globe, and one result of cinema's globalization has been a dramatic increase in regional film making, in local languages, with Japan's Ghibli studio a major beneficiary.

Film has been a global product since the early 1900s, when the French controlled the world market. Back then, as Kristin Thompson has shown in *Exporting Entertainment,* Pathé Frères dominated the U.S. market, where it released 50 to 70 percent of all new films. Financed by Jean Neyret and the Crédit Lyonnais bank, Pathé produced a film a day in 1906, dwarfing the combined output of all U.S. competition. It set up industrial, assembly-line production facilities in Paris and nearby Join-ville, permanently employing directors and actors. By 1908, as David Putt-nam writes in *Movies and Money,* Pathé's domination of world cinema was complete.

> He was selling twice as many films in the United States as all the American companies put together. Subsidiaries were producing films as far afield as Rome and Moscow, the latter accounting for half the movies in Russia by 1910. Pathé himself was a figure of such national prominence that when the actor Harold Lloyd wrote to him he simply addressed his letter to "Monsieur Charles Pathé, France." The second most powerful film company was Gaumont, also French and backed by the predecessor of Credit Commercial de France.[15]

In the early era, only French film makers had the backing of their nation's banks. U.S. film makers were so busy suing each other over patent and license infringements that U.S. banks would not loan them a shoeshine. But the U.S. market was changing in a way that would reorganize the industry: nickelodeons were becoming popular with the masses of immigrants in the cities. Nickelodeons screened twelve to eighteen programs a day, seven days a week. It made sense for them to have fixed places for exhibition and to *rent* films, so U.S. exhibitors who had been traveling around began to settle down, to attract a mass audience, and to improve the *logistics* of the business. As urban audiences grew, U.S. distributors bootlegged everyone's films, and exhibitors began to prosper.

In 1908, in the wake of suits by Thomas Edison and a succession of licensing deals by distributors, two quasi cartels arose, one mostly American (the Motion Picture Patents Company, or MPPC), the other mostly foreign (the Sales Corporation). They established uniform licensing fees, admissions, and import quotas. Edison invited Pathé to join his coalition, which allowed the MPPC to quash other foreign competition. Some American films were already being exported, with Britain as their largest market. Thompson notes that London alone had three hundred movie theaters in 1911. The MPPC started buying up film exchanges. This focus on distribution proved important: the 70 percent of the U.S. market that foreign firms had controlled in 1908 fell to 35 percent by 1910. In retrospect, as Puttnam points out, there were two keys to the American success in world film: U.S. firms dominated their home market, and they controlled distribution.

Film has a special economics. More than any other product of its day, the cost of a film—actors, equipment, sets—went into making a single item, the first print. The second and subsequent prints were comparatively cheap, anticipating the economics of many modern media products. The income they made was pure profit. Once clear, the economics of film distribution caused a revolution in the business, and by the end of 1912 American film companies accounted for 80 percent of new releases in their home market, as Puttnam notes, which they never ceased to dominate. Vitagraph, the third largest U.S. company, went a step farther, producing more prints at its Paris plant by 1912 than at its U.S. plant. Already it saw the perfect use for the "second print"—foreign distribution.

THE HISTORY OF U.S. FILM EXPORTS

World War I was the best thing that could have happened to U.S. film. It disrupted the powerful French, German, and Italian film industries, but left U.S. producers and the large Anglo-American audience relatively unscathed. The United States began producing narrative-driven film spectacles such as D. W. Griffith's *The Birth of a Nation* (1915), which drove up production costs 1,000 percent between 1910 and 1920. Stars such as Douglas Fairbanks and Mary Pickford pioneered on and off screen life-styles that emphasized personal freedom and consumption. By 1920 three times as many Americans per capita went to the movies every week as did citizens of any other country. Only 7 percent of Frenchmen saw a film a week. As John Izod argues, the huge U.S. audience, the high cost of film making, and the "aura" of American freedom and wealth gave U.S. film makers an edge.

After the war many European countries owed large debts to the United States, and importing film helped their repayments. American bankers, who had opened offices abroad to handle reparations and reconstruction, now looked more favorably on loans to an industry patronized by their wives and children. Given the low cost of the "second print," American firms could sell in this market at prices lower than even healthy European producers could have met.

The war also made New York City into the center of film shipping. America's late entry into the war meant that U.S. ships served Latin American, Asian, and even European markets without peril. By 1919 U.S. distributors had systems of finance and distribution in place that gave them worldwide scope. According to John Izod, exports of U.S. film rose from 36 million feet in 1915 to 235 million feet by 1925, while imports of foreign films to the United States declined from 16 million feet in 1913 to only 7 million feet in 1925.

TECHNICAL INNOVATIONS, LICENSING BATTLES, QUOTAS

A high level of technical innovation has spurred the export of many U.S. leisure goods. The introduction of sound film in 1927 was signal in this regard. Movie making became so expensive that the number of U.S. films made fell by 30 percent between 1927 and 1930, and most foreign film makers were not willing to take similar risks. At first, attendance at sound films soared, and exports rose dramatically. Louis B. Mayer in 1928 naively declared that sound film would lead to the universal use of English. But the novelty soon wore off, and Thompson notes that foreign audiences began to lose interest. It became clear that, instead of universal use of English, sound film created a language barrier and indicated the need for language markets. Only the largest—English, French, German, and Spanish—had economies of scale. In the rest there were additional costs for dubbing the second print or for subtitling services. Both of these added to second-print costs, though not enough to level the playing field for foreign producers. The U.S. producers temporarily captured 95 percent of the British, 75 percent of the French, and 68 percent of the Italian markets, according to Thomas Guback. Seen in this light, the introduction of sound film foretold the current situation: the dominance of Hollywood film is old news.

The advent of the Technicolor musical with *The Gold Diggers of Broadway* (1930) upped the technological ante again and led U.S. producers into the music business. Studios competed fiercely for viewers during the Depression, trumping each other with such innovations as crime films, sound effects, and deep-focus lenses. Foreign film makers could match neither the technique nor the spectacle—their national audiences flocked to the big-budget American imports. In markets as disparate as Japan, India, and China, 90 percent of the films shown in the 1920s were American.

Why couldn't European film makers compete? They faced their own blizzard of lawsuits over patents, some three hundred in Germany alone. French companies, which held few patents, concentrated on production and on building chains of sound-equipped theaters. Finally a Pan-European conference solved the patent dispute and set up a cartel known as Film Europe. One by one, European nations set up quotas on the number of imported U.S. films, as Thompson details. But the "quota wars" were ultimately for naught. U.S. firms opened foreign subsidiaries; Paramount and MGM went so far as to loan Germany's UFA group money to survive, gaining a guaranteed outlet in the joint subsidiary Parufamet.

Hollywood also began product differentiation. MGM had brilliant, plush interiors and a high-key lighting style. Warner became known for urban realism and low production costs. Disney pioneered cartoons. U.S. bankers reviewed all of their production plans. A. P. Giannini, eventual head of Bank of America, used funds from San Francisco's Italian community to invest in films, and Deutsche Bank, Crédit Lyonnais, and Crédit Commercial of France invested in Hollywood; in fact, since 1930 European banks have continuously invested in and profited from Hollywood film.

WORLD WAR II

During World War II, U.S. film was shut out of Europe. By 1940, Izod writes, it was distributed only in Switzerland, Sweden, and Portugal. But the moviegoing American audience and exports to Latin America and Asia saved it. The U.S. government also persuaded Hollywood to begin shipping newsreels to these markets to counter Axis propaganda, and by 1943 the cooperation turned into pro-American war films. After the war, film exports were part of aid to some countries under the Marshall Plan. Although there were censorship mechanisms, and a desire to propagandize for "the American way" in some quarters of U.S. government, actual distribution was too diverse to control. There was a hunger for Hollywood film in markets that had become addicted, such as Reinhold Wagnleitner describes in Austria in *Coca-Colonization and the Cold War* (1994). The studios finagled until they could export what they wanted. Rather than propagandize, they cleansed their films of material offensive to foreign audiences and added foreign scenes. They inundated France and Germany, offering films so cheaply that local producers could not compete. Europeans fell back on quotas, but again quotas failed. Not only did foreign audiences want the U.S. product, but the U.S. studios again set up foreign production companies, which took advantage of cheaper locations and labor to become "multinationals" before the word was invented.

THE MODERN ERA

Before World War II eight out of ten U.S. films recouped their production costs at home. After the war only one in ten could do so. By the mid-1950s some 40 percent of ticket revenue for U.S. film was already coming from overseas, a figure that increased to 53 percent by the early 1960s. Izod gives figures showing that Europeans were the source of 80 percent of the new revenue, with Japan joining them after its occupation. So film producers began to look for international appeal: if a foreign scene was possible, they added it. But the biggest change was "runaway production." In order to meet foreign quotas and limits on repatriation of profit, U.S. producers began shooting some films overseas. In 1949 U.S. studios produced only 19 films abroad, but by 1969 that number reached 183 films. Beginning in the 1950s, Puttnam points out, U.S. producers also sold films to television, while European film makers boycotted this market. France proved the market most resistant to American film imports. Attendance in France was far lower than in other European countries, and French exhibitors refused to build theaters in small towns, because it was not a "sure" investment and because the Catholic Church objected. Thus in 1960 French films still accounted for more than 50 percent of French box office receipts, and the government was pouring one-half billion francs a year into French film. "Despite this unprecedented level of aid," writes Puttnam, "the fastest growing French genre of the 1960s, in numerical terms, was not Nouvelle Vague films, but pornography, and that story was repeated across the rest of Europe." The top grossing film of 1974 in France was the soft-core *Emmanuelle*. During the 1970s, Puttnam adds, "pornographic films accounted for an average of 50 percent of all French productions."[16]

Meanwhile Americans were building "platforms," such as *The Godfather* (Paramount, 1972, 1974, 1990) and *Star Wars* (Twentieth Century Fox, 1977, 1980, 1983, 1999, 2002, and 2005). From these, they launched sequels and products like toys and books. Videotape rentals were another American innovation of the 1970s that became a major source of profit. *Star Wars* earned $100 million in videotape rentals, and other films that failed in theaters made a profit in rental stores. By 1995, John W. Cones writes, rentals had "outpaced theatrical revenues [box office] both in the foreign and domestic market for several years." In that year, he notes, film executive Richard Childs said that "domestic gross sales of all home videos are about twice that of box-office gross."[17]

Foreign presales were another innovation. Dino De Laurentis, a small-time producer in the south of Italy, who should have been devastated by World War II, managed to produce the hits *Bitter Rice* (1950) and *La Strada* (1951). Ben Wasser writes that De Laurentis used his collaborations with Federico Fellini and Carlo Ponti to presell their films in the United States. In 1972 he came to Hollywood, where he produced several films in the following manner: he obtained an Italian government production subvention; he presold North American rights for about 50 percent of needed funding; he presold world distribution rights; and, finally, he sold off the whole project bit by bit to investors as he actually filmed it. As one Walter Reade studio executive exclaimed, "Dino is the only producer who thinks of the United States as just another territory."[18]

Soon Americans were following Dino's formula, and by 1973 their overseas sales reached $415 million, exceeding their domestic sales. De Laurentis's next innovation was to get preproduction funding from his overseas distributors, a process that was expedited by Frans Afman of the Slavenburg Bank of Rotterdam. Based on the distributors' written guarantees, Afman and the bank advanced production money to De Laurentis for filming, usually mass-market fare. But unexpectedly one of their films, called *Death Wish* and starring Charles Bronson, turned out to be an enormous hit. Not only did De Laurentis look like a genius, but his finance method became the standard for Hollywood.

Crédit Lyonnais, by 1982 a French-government-owned bank, claimed to be the number one bank in film finance in the 1980s. It had deals with Carolco, Cannon, Castle Rock, Cinergi, De Laurentis, Empire, Epic, Hemdale, Largo, Morgan Creek, Nelson Entertainment, New World, Sovereign, and Trans World Entertainment. During the socialist Mitterand era, this government bank backed an A-list of "Hollywood" hits: *Dances with Wolves, Crimes of the Heart, Blue Velvet, Salvador, Platoon, Hoosiers, A Room with a View, Superman I,* and *Superman II.* "This French bank can be credited with helping the proliferation of new studios in the early eighties," writes Wasser, "by [its] willingness to underwrite global pre-sales."[19] Today the French complain about unfair competition, but in the 1980s they also owned and operated MGM, even appointing the entertainment division executives.

There are other examples of foreign control in Hollywood. In the 1990s Sony took over Columbia, then Matsushita bought MCA/Universal. Sony was particularly aggressive, battling with the much smaller MGM and claiming rights to James Bond and Spiderman to which it had no clear title. By October 2002 Sony was the world's leading film producer and distributor in revenues ($1.4 billion) and market share (19.4%).[20] Then Australian Rupert Murdoch purchased Fox, expanding his global entertainment empire. Other foreign players are France's Canal+, Germany's Cappella, and Japan's Largo Corporation. Universal was purchased by Seagram's (Canada), then by Vivendi (France). Vivendi's Canal+ promoted Universal's *The Hulk* on its twenty cable channels in sixteen countries. By late 2002, three of the top five Hollywood distributors were foreign. They controlled 41 percent of the world market: Sony 19.4 percent, Twentieth Century Fox 11.8 percent, and Universal 9.8 percent. Then in 2004 General Electric bought 80 percent of Universal. It became hard to keep track.

The typical "Hollywood" blockbuster in this era was something like Fox's *X-Men* (2000), based on a comic book. It starred an Australian (Hugh Jackman), a couple of Brits (Ian McKellen, Patrick Stewart), and some exotic Americans (Halle Berry, Rebecca Romijn-Stamos). It was filmed in Canada. Seven production companies were involved, four of them American. There were eleven distributors, nine of them subsidiaries of Australia's Fox empire that dealt with the French, Spanish, Japanese, German, Italian, and Portuguese markets. Fifteen U.S. special-effects companies were employed, but they outsourced work to Taiwan, Spain, Mexico, and the Philippines.

Due to the tax laws, quotas, and repatriation rules that affected *X-Men,* Fox could afford to import *Bend It Like Beckham* (2002). This soccer film was produced by five British, one American, and four German firms. Indian-British director Gurinder Chadha had no hits when Fox put her film on 555 American screens. By May 2003 her $6 million film had grossed more than $100 million, not counting resales, video rentals, or the DVDs (available in three languages) and audiotapes of the music (available at Amazon.com) by various Sikh and Punjabi artists, who also received royalties.

CINEMA IS LOCAL

Given its proximity, Mexico would seem vulnerable to colonization by Hollywood film. It shares a two-thousand-mile border with the United States, lies within reach of American television and radio transmitters, and has millions of citizens who visit the United States. But moviegoing is different in Mexico, for Mexicans choose films within their own cultural matrix.

Back in the nickelodeon era, the United States supplied 50 percent of Mexico's viewing. In the era of silent film, U.S. consuls reported, Mexicans preferred French films and a few U.S. westerns. During World War I the disruption of shipping made delivery of French films undependable. But Thompson writes that

> due to the civil war there, both economic problems and the difficulty of shipping kept American firms away. By early 1916 Mexico's currency had dropped off to less than one-fifth of its pre-war strength in relation to the American dollar. . . . After the USA went into World War I, the American government placed very strict limitations on exports of films to Mexico and it was not until 1919 that conditions settled down enough to make it a viable market.[21]

After the war, U.S. propaganda films (*Ford Factory, Ford Tractor*) failed as badly in Mexico as elsewhere. Mexico, which had supplied actors and locations for hundreds of cowboy movies, provided a discerning audience. Strife-weary viewers wanted Charlie Chaplin, Norma Talmadge, Mary Pickford, and Douglas Fairbanks.

> In 1921, the Mexican market was finally becoming stable enough to be attractive to American firms, which began opening offices in Mexico City. But early in 1922, the government there placed a ban on films coming from any company that made movies portraying Mexicans in an offensive way—a ban that applied even if the offending films were not themselves sent into Mexico. Some American companies promised to

comply with the Mexican demands. By the summer of the next year, an American official in Yucatan reported American films "by far the most popular." Import figures for the first five months of 1925 show the American market share at a level comparable to those in South American markets.[22]

Bending to local taste, Hollywood began to capture the Mexican market. The advent of sound increased U.S. dominance, as Mexico became the location for Spanish-language versions of American films, with Mexican actors playing major roles. While this eventually proved limiting (Spanish speakers in other markets took offense at the Mexican accent), Mexicans appreciated it, and many Mexican actors became bilingual. When the Spanish-language units folded, actresses such as Dolores del Rio went straight to Hollywood. By 1930 the United States had a 98 percent share of the Mexican film market. This stayed constant until the beginning of World War II, writes Thompson, when it dipped to 90 percent. But today the number of Hollywood films shown in Mexico is much lower. Here's why.

A TRIP TO THE MEXICAN CINEMA

The consumption of film follows local habits. Unlike France, almost no Mexican town is too small for a cinema, though the films may be saccharine or blow-'em ups, but multiplexes are becoming common, especially outside of urban centers. Screening rooms tend to be large, even in new theaters, rather than the boutique models of the United States and Europe. First-run cinemas are often grand affairs, with marble floors and grand staircases leading to the balcony seating. The *cine* page of the Sunday edition of Mexico City's *Reforma* listed 187 cinemas in 1995, of which about 80 were in fourteen multiplexes of four or more screens.

Crowds at the movies are more reserved than in the United States. There is less talking, and none at all during showings. There are more viewers over age twenty-five and more couples than in the United States. Except for *niños fresas* in their neighborhood theaters, groups of teenagers are not common. Attire is dressier than in the United States—no shorts, sandals, halter tops, or muscle shirts. High heels and ties are common. Patrons seem to consume twice as much popcorn, called *palmas* or *palmettos,* and soft drinks and candy as in the United States. Few trash receptacles exist, but theaters are well cleaned between showings. The most unusual feature of Mexican cinema is the "intermission." Halfway through each film, without warning, the film cuts to a commercial for Snickers or Coke or *palomas,* exhorting the audience to charge down to the snack bar.

The ingredients of film success are different too. Let's take a week of May 1995 as a snapshot. For a city of 15 million people, there is relatively little variety—just twenty-eight films are playing in Mexico City. As the week begins, nineteen are American made (68%), but as the week ends only fourteen are U.S. made (50%). Two of the "American" films are co-productions with other countries, such as *Prêt-à-Porter* (United States–France). Mexico itself accounts for five films as the week begins and six as it ends, with a Mexican-Spanish-Cuban coproduction raising the number to seven. Spain and France have two films each as the week starts, one each as it ends, while Russia and Poland each contribute one film during the period.

The American-made films are showing on 105 screens as the week starts but on only 90 screens as it ends. Mexican films are shown on only 19 screens on Monday but on 23 screens the next Sunday, the result of three new films (*Callejón, Guerrero,* and *Me Tengo*) that each add one or two screens. Meanwhile second-run imports, such as the U.S.-made *Mujercitas* (*Little Women,* 1994) move around the city to run alongside bigger hits such as *Epidemia* (*Outbreak,* United States, 1995) and the Spanish *Kika* (1993). Other nations account for 32 screens. About 20 cinemas are not accounted for, at some point in the week. Possibly these are rerun houses, in the midst of changing titles, or undergoing renovations. Overall it appears that the United States is the source of 75 percent of films on first-run screens.

The films playing on the most screens as the week begins are *Epidemia* with twenty-five and *Street Fighter: La Ultima Batalla* (United States, 1994) with twenty-six. When the week ends, *Epidemia* has spread to thirty-one cinemas, while *Street Fighter* retreats to twenty-four cinemas. The biggest changes over the week will be the opening of *Los Caprichos de la Moda* (*Prêt-à-Porter,* France–United States, 1994) in twenty-two cinemas, along with Jackie Chan's *Un Loco en Hong Kong* (*Shuang long hui,* Hong Kong, 1992) and *La Noche del Demonio* (*Tales from the Crypt: Demon Night,* United States, 1995), playing in twelve and twenty-two locations, respectively. The four Mexican films are playing in a number of first-run houses, as is *Kika,* a Spanish comedy by Pedro Almodóvar (fifteen screens).

These films represent an array of bets made by distributors and chains on the roulette table of spectatorship. Some films seem sure hits, such as

Street Fighter, since there is always an audience for violent action films. Some, such as *Prêt-à-Porter* (which will bomb), represent star power and contractual obligations. Others are the biggest hits of the previous year in the United States: *Pulp Fiction, Forrest Gump,* and *Dumb and Dumber* (they also bomb).

The Mexican films have mixed success. The screwball marriage comedy *Me Tengo que Casar* (dir. Manuel García Muñoz, 1995) draws bigger audiences than the ambitious *El Callejón de los Milagros,* which is a paradigm of the "international" film that government film boards love. Audiences like the *telenovela* romantic formula of *Me Tengo,* while *El Callejón* has unclear messages about homosexuality, marijuana, and migration to the United States. Whereas the first is funny and familiar, there's a sense to the second that a government committee decided what issues would be addressed and how deep the examination could go. More popular than either film is *Fresa y Chocolate* (*Strawberries and Chocolate,* Cuba-Mexico-Spain, 1993), which shows at only one theater but to very big crowds.

The big successes among U.S. films are *Epidemia* (*Outbreak*), up from twenty-five to thirty-one cinemas at week's end, and *Rapa Nui,* a Kevin Costner–produced epic that failed miserably in the United States. *Street Fighter* is also doing a good business. It is ostensibly an American film, but reading the credits we find a runaway production. This Jean-Claude van Damme sequel is cofinanced by the United States and Japan and filmed in Australia and Canada. Investors got their return: it grossed $33 million in the United States and $70 million worldwide by 2003. None of these "American" films got good reviews in the Mexican press.

But Mexicans don't follow critical reviews, to judge from the ratings in *Reforma.* It gave *Todas las Mañanas del Mundo* (*Tout les Matins du Monde*) four stars, its highest rating, and *Tiempos Violentos* (*Pulp Fiction*) three and a half, yet each played on only a single screen and drew sparse audiences. *El Callejón del los Milagros* and *Fresa y Chocolate* rated three stars (good) but played on only nine screens. A much bigger hit with the public and in terms of screens was *Epidemia,* which rated two stars ("ordinary"). Mexican exhibitors don't quote reviews or reviewers in their ads, but word of mouth seems to spread very fast.

Judging the ingredients of success or failure in a film is difficult in a foreign country. The mechanics of subtitling or dubbing are important. Like any audience, Mexicans like to hear their language spoken, in their own accent. Subtitles are a chore to read, and in a nation where the literacy rate is low, many moviegoers don't read quickly. *Rapa Nui* had huge subtitles that sometimes occupied up to one-quarter of the screen and blocked the action, but they were very clear and stayed on screen for a long time. On the other hand, *Prêt-à-Porter* had the discrete, white, bottom of the screen subtitles that Western audiences expect of an art film; they were long, sometimes pointless, often badly translated, and did not stay up long. Cultural factors play an important role in the reception of dialogue. Imagine trying to write subtitles for *Pulp Fiction*, which gained its cachet in the United States by skillful parody of the vacuous dialogue in pulp detective novels. Of the twenty-three flavors of "fuck" in Mexican Spanish, which best replicates John Travolta's favorite word? Even if the film had been "translatable," it probably would have failed because it depended so heavily on the American pulp novel tradition, with which Mexicans are only vaguely familiar.

Dialogue and subtitling were also limiting factors for *Forrest Gump* and *Dumb and Dumber.* The mental peculiarities of Forrest Gump, and the peculiar flatness of Tom Hanks's voice as he communicated them, do not come across in subtitles. Just as no American could tell by his accent that a Spanish speaker was mentally handicapped, or be amused by the culturally corrective power of this shortcoming, no Mexican could be expected to pick up on the same qualities in an American actor. The *loco* vacuity of *Dumb and Dumber* is similarly dependent on tonal qualities in the actors' delivery. Both films depended on nuances of verbal expression and the audience knowing what was inappropriate or maladroit, which could be appreciated only by native or near-native speakers.

Epidemia and *Rapa Nui,* in contrast, did not depend on dialogue. Both were chock-full of action and plot. Much of this was "action" in the most clichéd Hollywood sense—helicopters and tanks, love plots, special effects—but the narrative was always clear. In *Epidemia* Dustin Hoffman and Rene Russo are "federal doctors" (*medicos federales* in the subtitles) who fight to save an entire town that may be wiped out by a deadly virus. Even though the hamlet on the Oregon coast with its cream-complexioned residents looked as foreign as Mars to most Mexicans, the object of the plot was immediately clear. Viruses such as the deadly ebola are not merely the stuff of film in Mexico. Typhoid fever, hepatitis A, and dengue are a reality in the Chiapas, Oaxaca, Campeche, and Veracruz departments. No one in the audience found it ironic that during the "intermission" two government public service announcements instructed patrons on proper hand washing and sanitation measures to combat cholera and malaria.

A movie about disease afflicting those richer, seemingly omnipotent neighbors to the north is intrinsically interesting. "Try to stay calm!"

warns the film's advertising. If the plot joins a deadly disease with questions about the intent of a monolithic, militarized government (the PRI still ruled Mexico in 1995), then differences in race, class, residence, and language can be put aside. The core issue was compelling in Mexico, and there were enough action, special effects, and romantic interest to carry the audience. *Epidemia* did a good business.

So did *Rapa Nui,* a film pummeled by critics and the box office in the United States. The film depicts a Stone Age tribe, organized by competing clans, living on an ecologically limited Easter Island kind of locale. The clans deplete their resources to raise stone heads to the god they believe will come to their rescue. Mexican audiences appeared powerfully affected by two aspects of the film. The enslavement of the clans by the priestly caste to raise the useless monuments drew comments of recognition all around the theater. Although it's not easy to say who Mexicans saw as the priestly caste (Aztec priests? the PRI? the church? the United States?), they expressed a sense of being pitted against each other by occupation, by region, and by ethnic group (Europeans vs. *mestizos* vs. *indigenos*). Most working-class Mexicans think this unfortunate. They would like to make common cause against this "priestly caste."

Like the Easter Islanders, Mexicans live surrounded by huge stone artifacts raised to appease the gods. From the Templo Mayor under Mexico City's main square to the Toltec Pyramids of the Sun and Moon at nearby Teotihuacan, they see monuments whose cost in human sweat and blood can only be estimated. The ecological price of this monument building is only too clear, for Mexico suffers the same irrational deforestation depicted in the film. Citizens are aware of their ecological crisis, and they were on the edges of their seats awaiting the resolution of this film, which unfortunately showed Jason Scott Lee sailing off to a land over the horizon where the mysterious technology of iron exists. For Mexicans this translates to yet one more version of flight from home to the Great American Paradise, but *Rapa Nui* played to full houses.

On the other hand *Los Capriccios de la Moda* (*Prêt-à-Porter,* France–United States 1994) played to fifteen people the night I saw it, in a first-run cinema with marble floors, a chandeliered balcony, and three-dollar (expensive) tickets. Five people left in the first forty minutes. Paris? The fashion world? Robert Altman says it is shallow and empty and nepotistic—isn't that obvious? The audience liked the low comic scenes of the parallel romances between Sophia Loren and Marcello Mastroianni, and Tim Robbins and Julia Roberts. The critique of the "fashion system" they found utterly pointless. By the end of the first week, the newspaper ads were changed to read "31 film stars, 9 sensual couples, 14 naked super-models, 1 musical hit."[23] And one big loss.

The Mexican share of the Mexican market could be bigger, because the government subsidizes film production and Mexico has scads of good directors. In the 1940s and 1950s Mexico cranked out melodramas and comedies for all of Latin America. The problem is that the government wants to fund only European-looking films, to find the Carlos Sauras and Pedro Almodóvars of Latin America, rather than to play to traditional Mexican market strengths or to export to the Latin market. As Emilio Garcia Riera said, "the Mexican filmmakers are excellent, but the industry has produced only trivia."[24]

Callejón de los Milagros (dir. Jorge Fons, 1995; *Miracle Alley* in English) is an example. Based on Egyptian Nobel laureate Naguib Mafouz's novel, coproduced with Spain and Cuba, responsible in its view of social issues, and featuring Mexico's best young "serious" actors (Salma Hayek's first international role), *Callejón* managed to clear all the committee hurdles. It received eighteen nominations for the Ariel (Mexico's Oscar), but to Mexican audiences the film still smacked of *cinema de lágrimas* (the tear-jerker), a genre that Mexican TV studios churn out better. This genre has been a training ground for many film makers, but when one adds to it *Rashomon*-like multiple viewpoints, the struggle to come to terms with sexuality, and artificially lit sets, the result has no genuinely local charac-ter. As one critic wrote, "This film is just one banal art-house cliché after another."[25] And it exported poorly, unlike the later hits *Amores Perros* and *Y Tu Mama También*.

WHO MAKES MONEY IN FILM?

It is commonly assumed that "studios" receive the lion's share of Holly-wood film's profit. This is half true. "Studios" today are simply investors who may do very well on one film, while losing heavily on three or four. Despite starring Harrison Ford and Brad Pitt, *The Devil's Own* (1997) lost $40 million. The actors made big money, but the investors lost their shirts.

Unlike actors, investors take big risks. The first investors were out-siders, often *arriviste* entrepreneurs, nickelodeon kings like the Warner brothers of Pittsburgh. Then venture capital firms, such as Kuhn, Loeb and Goldman, Sachs, who had credit with East Coast banks, became in-vestors as theaters were set up. When the business became stable, the banks themselves followed, including foreign ones like Crédit Lyonnais, as did investment houses. These three layers of film finance are still com-mon. Thousands of "outsiders" hope to strike it rich, converting Texas or

Saudi oil money into Hollywood gold. With their money from Brunei or Colombia or Russia, they back "Hollywood" films. The risk is high, but these people also invest in racehorses and oil wells. Their lawyers and accountants have created tax shelters that allow them to begin to write off their investments before a film is even finished.

With the first investors' money, the producers draw in more established sources—banks—on the basis of their previous hits. Retirement funds, for example, may participate in a film at this point. Many of them invest not only in studios such as Disney, Sony, and Time Warner, but in distribution and in exhibition, because Regal, Loews, Cinemark, and AMC are good stocks. These exhibitors indirectly finance film by advancing funds or signing prepaid contracts for screenings. Then, as John Cones shows in *The Feature Film Distribution Deal,* comes the linchpin, the distribution deal. Everything runs through the "distribution deal"—book rights, advertising, "anticipated expenses," "below-the-line fringes," "over-reported travel expenses," "exhibitor expenses," and so on. There are a dozen different kinds of "interest expenses" on the monies loaned to the film-in-the-making, and there are lucrative resale markets in television, syndication, overseas exhibition, and videotape or DVD. Distribution is where money is made and is what people fight about.

Entering the actual film distribution business today would be like opening a gas station to compete with Exxon. Ever since they drove the exchanges out of business in the 1920s, distributors have controlled the film business. They have worn down, evaded, and defeated quotas. The big distributors are names that we know—Fox, Buena Vista, Sony, Disney, Warners, Universal Paramount, Columbia—and some less famous: Canal+ (France), Eros International (India), and Canyon Cinema (independent). Distribution is a business that profits even more than before from economies of scale and logistics, because exhibitors depend on the unquestioned, punctual delivery (and pickup) of films, so that having a system in place in every large city is crucial. Distributors are almost invisible, and they may run their operations from Chicago or Hong Kong. They are the paradigm of logistical expertise, the subject of chapter 3. They make more films available to more viewers in more countries, more cheaply, than anyone dreamed possible in 1900.

But the film business is still local in an important aspect: exhibition. And the United States is the most important exhibition locale, because Americans are still the most moviegoing people on Earth, seeing five times as many films per capita a year as the French. When bargain nights and passes are figured in, U.S. prices are the lowest in the developed

world. But in most accounts of Hollywood's "hegemony," the local economics of exhibition go unexamined.

To understand how much film money remains at the local level, let's
look at Cleveland, Ohio. It has thirty-four movie venues, most of them
multiplexes, with 380 screens. The Regal chain accounts for over 100
screens, followed by Cinemark (45), but local entrepreneurs own the third
(Cleveland Cinemas, 39 screens) and fourth largest groups (Atlas, 32
screens). They are followed by AMC (21), Loews (20), and Magic Johnson (12). There are also niche markets, ranging from drive-ins and second-run movies to art house and Omnimax. It is impossible to say how
profitable the average film is in Cleveland, but probably 60 percent of the
average $7.50 ticket in 2003 stayed in Cleveland. It paid the theater's mort-
gage, electricity, personnel, and advertising. The theater and its employees also paid a variety of local taxes. The owner's profits were also taxed
locally and perhaps reinvested locally. Most cinemas were hubs for restaurants and bars, or for shops in malls, which depended on them to generate an "evening out." These enterprises in turn paid rents, employees, and
taxes.

Considered in this economic matrix, the amount of profit exported to
the "studios" is relatively small. In most nations the situation is similar.
Walking the streets of Paris, we see the Gaumont and Pathé chains. The
Pathé theaters may be showing *The Hulk*—after all, Vivendi produced it—
but they are paying French employees, mortgages, and taxes, and they are
buying French newspaper and television advertising and French conces-
sion supplies. On the Avenue des Gobelins near Place d'Italie, where I usually stay, three French-owned multiplexes generate the "evening out" for
more than fifty restaurants and bars, all French owned.

If we travel back to Nishinomiya again, filmgoing is also local and
different from Mexico, France, or the United States. We have to take a
train into Osaka to see a movie. The filmgoers there are young, urban,
affluent, and willing to stand in line for hours to see a first-run film.
Women and teenagers dominate; the couples are on dates or in groups.
"In the United States, it's not unusual for a married couple to go to the
movies or a show after work," says a Toho chain executive, but "in Japan
if a man told his friends he was going home early to take his wife to the
movies, they'd say he's weird."[26] There is no elder audience in Japan, nor
parking. Theaters are smack downtown, close to restaurants and bars
and the train stations. Filmgoing is expensive, often planned several days
in advance. Attending the film is, like everything in Japan, an exercise in
punctuality, courtesy, cleanliness, and appropriate group response. Tick-

ets are for assigned, numbered seats. *Spirited Away* (*Chihiro*, Miyazaki, 2001) was more than a movie in Japan: it reaffirmed a cultural vision of these qualities. New in Japan is the multiplex, which exists only on the outskirts or in rural areas, where land is cheaper. Chain operators Toho and Shichiku eschew multiple screens, preferring to dribble the hits out one at a time and milk the market.

While India makes the most films in the world (1,200 in 2002) and the United States is second (543), Japan is a surprising third (293). It makes more movies than France and China combined. Some of these, such as the masterpieces of Hayao Miyazaki, are distributed worldwide. Others, including the highly profitable *anime,* are distributed only in Asia. Then there are samurai, *yakuza* (gangster), and romance films that are culturally specific Japanese genres, which hardly export at all. All of these make far more money in videotape and DVD rentals in Japan than they do in screenings, because the Japanese watch a huge amount of film at home.

If we enter one of the two rental shops by the Kitaguchi station, we see a bustling business, even at midday. There are thousands of titles, updated daily. And there are still multiple copies of *Titanic* (1998) available. "*Titanic* played very well in America," writes Lynn Hirschberg, "but in Japan they loved it, loved it, loved it. It eventually made $900 million in movie rentals worldwide." The mechanism of this phenomenon was identified by James Ulmer, who operates a website that ranks the cachet of actors with audiences: Japanese women adored Leonardo DiCaprio. They went to *Titanic* on bargain day with their girlfriends, then dragged their boyfriends on Saturday night, then rented the film (often several times), and finally bought it. According to Ulmer, they have also driven the careers of Brad Pitt, George Clooney, Hugh Jackman, and Orlando Bloom, as well as many Japanese stars.[27] Even as we are standing in the rental shop, a woman in an "office lady" uniform picks *Titanic* off the shelf and takes it to the desk, a look of remembered satisfaction on her face.

SUBTITLING AND DUBBING

The local economics of film extend to subtitling and dubbing. The lucrative Japanese, German, French, and Italian markets demand dubbing, which is the most expensive way to adapt a film. Scripts have to be translated (by a local translation company). Local actors have to be hired, at union rates, and rehearsed. There are even dubbing directors. Local studios are rented, and local recording engineers employed at union rates. The new dialogue is then remixed with the original sound effects and music, also in local facilities. Today there are film-dubbing industries in the eleven largest non-English-language markets, plus Italian, Polish, Greek, and Serbo-Croatian. If Hollywood films were not imported, these jobs would not exist. A 1998 strike by unionized dubbers in Italy, where eighty dubbing companies employ more than a thousand dubbers, shut down the nation's theaters, some television broadcasts, and some commercials. Italian dubbers have their own Oscar (called the "David"), and some, like Ferruccio Amendola (the voice of Robert De Niro, Sylvester Stallone, and Dustin Hoffman) receive royalties and appearance fees. "Italians aren't used to seeing movies in the original version," says Richard Borg of United International Pictures, "They just won't go."[28]

Subtitling is less expensive but hardly free. It is used in two types of market. One consists of educated, urban moviegoers in France, Germany, or the United States; the other markets such as the Cebúano audience of the Philippines or the Hausa audience of Nigeria. For the first group, interested in the artistic integrity of the original, subtitling has to be more accurate and deft than dubbing. Literary translators are employed. Adding subtitles to these films is more difficult than it appears, because dark backgrounds are not always available at the bottom of the screen at the right moment. Subtitles also make more sense than dubbing for the Cebúano or Hausa audience, because exhibitors would lose money showing a dubbed version to an audience numbering only in the thousands. Subtitling is another part of the ticket price that stays at the local level.

RUNAWAY PRODUCTION

The local economics of "Hollywood film" are most evident in the vast sums spent on "runaway production." A $50 million budget film, shot in Morocco, spends about $25 million in that country. De Laurentis pioneered this method and was so successful that Twentieth Century Fox and United Artists soon followed suit, making the Sergio Leone "spaghetti westerns," Fellini's *Satyricon*, and Bertolucci's *Last Tango in Paris* abroad. Most of this work was done at Rome's Cinecitta studios. Next, United Artists went to France, where it funded films by François Truffaut, Claude Lelouch, Louis Malle, Phillipe de Broca, and others. In Sweden it financed four films by Ingmar Bergman. In Great Britain United Artists funded the Beatles' *A Hard Day's Night* and the original James Bond series. Runaway production reached extraordinary levels in the 1980s and 1990s. *Schindler's List* was made on location in and around Cracow, Poland, where Steven Spielberg spent millions on local actors, cameramen, technicians, carpenters, translators, food, lodging, vehicles, costumes, and various rentals. Today some citizens of Cracow still earn a living giving "*Schind-*

ler's List tours," and the bleak former ghetto featured in the film has been transformed into a trendy nightlife area.

There is another advantage to runaway production. Many countries have laws limiting how much of the exhibition profit of a foreign film can be repatriated. The cultural politics of these laws would fill a book. In the short term, they seem to protect niche culture industries, but over the long haul they are ineffective against major cultural commodities, such as blockbusters. Major gaps in such laws allow runaway production and films produced by local subsidiaries of international studios (so they are not "imported" and their profits can pay for their costs). The consequence of such laws is that it now makes particular economic sense to film in lower-cost countries where the film will also show well, especially Australia, Great Britain, or Canada. In the case of *Schindler's List,* producers did not have to dream up ways of translating their zlotys into dollars; their Polish profits didn't begin to equal their Polish expenses, so as a money-losing "Polish" film company they qualified for a government subvention. By 2005 "Hollywood" films such as *The Matrix* series, *The Bourne Supremacy, Ocean's Twelve,* and *Troy* were routinely shot entirely overseas.

As a result of these changes, Hollywood's overseas box office receipts exceeded its domestic receipts for the first time in 1994. The globalization of capital, a result of deregulation of financial markets, encouraged foreigners to buy into "Hollywood," which became globally owned. "What appeared most surprising," writes Puttnam,

> was that companies from France . . . should be at the head of those lining up to invest in Hollywood . . . The management of conglomerates in the French film industry, as well as elsewhere, no longer felt obliged to defend those established values of high culture so long espoused by their intellectual compatriots. Whatever their underlying rationale may have been, companies like Canal+ and Chargeurs had no compunction about investing in Hollywood while at the same time the majority of French producers and policymakers were engaged in a strenuous battle to prevent American movies from flowing unimpeded onto Europe's cinema and television screens.[29]

Indeed, French intellectuals led the 1993 GATT battle over Europe's need for a "cultural exception" to free-trade laws in film and television. And they won—"a victory for art and artists over the commercialization of culture," said former French Minister of Culture Jack Lang.[30] In addition to the ticket surcharge, the French poured up to $400 million a year into government subventions for their own films.

But the enemy had vanished. By 2000 Hollywood was largely a place where deals were done, where studios, as Warner Brothers vice president Richard Fox said, were just "distributors, banks, and owners of intellectual copyrights, contracting out creative and production activities to others."[31] The profits landed in bank accounts in Riyadh or Vienna or Chicago, but the U.S. audience no longer determined what was made. In Japan, *Spirited Away* by Miyazaki drew an audience of 20 million and earned 30 trillion yen. That surpassed by 15 percent the record held by *Titanic* (1998), which had drawn an audience of 16.8 million and made 26 trillion yen. Local preferences ruled the Japanese market, and Japanese investors began to commercialize the export of *anime* to the United States. In France in 2001, "French films accounted for almost 50% of the total French box office."[32] Russians produced half of the films shown in Russia in 2002, and seventy-five films in 2003, as ticket sales increased tenfold over 1999, and as Russia's $175 million box office approached France's $185 million box office. Films were being made on budgets of $15,000 in Nigeria's "Nollywood," dominating that nation's $45 million home market. By 2003 Nigeria produced four hundred films a year, which were widely influential in other African nations and earned $45 million abroad. In late 2003 the number one film in Malaysia was not from the United States but from South Korea, which produced 50 percent of the films shown on its own screens in 2003, with ticket sales up five times over 1996. It earned $15 million on film exports. Film had become a world business in which the United States was just another competitor. "It is now less a matter of Hollywood corrupting the world than of the world corrupting Hollywood," wrote the *Economist* in 1998. "The more Hollywood becomes preoccupied by the global market, the more it produces generic blockbusters made to play as well in Pisa as in Peoria. Such films are driven by special effects that can be appreciated by people with a minimal grasp of English rather than by dialogue and plot. They eschew fine-grained cultural observation for generic subjects that anybody can identify with, regardless of national origins. There is nothing particularly American about boats crashing into icebergs or asteroids that threaten to obliterate human life."[33]

American Television and the Rise of Local Programming

Compared with other forms of popular culture, television moves at light speed, changing formats in a constant competitive search for novelty and audience. For some critics, this makes it a kind of fifth column. "It is enough to think of all the people who watch exported North American television programs to realize that this cultural intervention is deeper than anything known in earlier forms of colonization or imperialism," writes Fredric Jameson. "Television in some other (not merely third world) countries is almost wholly colonized by imported Northamerican shows."[34] Only if you are a member of what Peter Berger terms "faculty club international," staying at Hiltons overseas and watching cable, is it possible to believe that U.S. television programming rules the airwaves. Today television is regional or national, and powerfully so, as Al-Jazeera has shown.

A large English-language market, technical invention, and innovative financing propelled the initial export of American programs. That was understandable, because the United States led the world in the adoption of television, buying 36 million sets by 1955 (versus 4.5 million in Britain, only 300,000 in the rest of Europe, and a sprinkling in Japan at that date). When U.S. equipment makers agreed on the NTSC (National Television System Committee) standard and the Ampex Corporation in 1956 demonstrated magnetic tape recording, the export of U.S. programs became possible. After all, hardly anyone else was even making programs. But the initial export market was only in equipment. Foreigners needed sets first, and sets were expensive, several thousand each in current dollars. Critics like Jameson assume that today everyone has a set, but that's still not true: millions of Asians and South Americans do not, and fewer than one in twenty Africans owned a television in 2004.

CBS was first to establish a foreign program distribution subsidiary, followed by ABC and NBC. Learning from the film industry, they licensed programs from independent producers, which they syndicated at home and abroad. *Lassie* made $4 million in foreign revenue by 1958, according to scholar William Boddy, and five of the top ten shows in Japan that year were U.S. series or clones. By 1961 CBS Films was selling about fifteen hundred half-hour episodes in fifty-five countries: *The Lone Ranger* alone showed in twenty-four countries.

But what happened next indicates the general trend. Using their own technical innovations, such as Toshiba's recording system, the Japanese domesticated and then remade U.S. shows. They passed from copying to creating their own unique music, cooking, and comedy formats. Then

Japan used the 1964 Tokyo Olympics to ramp up its national infrastructure. People bought sets to watch the Olympics, and the government
financed state-of-the-art transmission facilities. By the 1970s Japan was a
net exporter not only of high-quality TV sets, in which it soon dominated
the American market, but of cartoon *anime* and the "funniest home
video" genre. It imported only hit dramas such as *Dallas* and *Dynasty*
(and later *Allie McBeal*).

PROGRAM DIVERSIFICATION

Just as movies had, television programming entered a period of product
differentiation. The introduction of UHF (ultra high frequency) channels in the 1960s and cable in the 1970s created many, many outlets in the
United States, which were on the air at all hours, creating an unquench-
able demand for programming. The large English-language market meant
that successful programs could be syndicated in the United States and
sold in other English-speaking markets, on the logic of film's "second
print." If very popular, they could be dubbed and sold in other language
markets. Exports to Japan had shown that the best bet was drama, prefer-
ably with high production values and known stars. A sporting event or a
music show might sell abroad once, but syndicated series developed
lucrative repeat audiences.

In contrast to the United States, Boddy points out, most foreign
nations subsidized their television systems. The BBC in Britain, CBC in
Canada, the RTF and ORTF in France, and ARD/ ZDF in Germany produced programming that had no secondary market. With outlets multi-
plying, they were then handicapped. The Nielson rating system appeared,
showing that viewers liked American drama, such as the miniseries *Roots*
(1977) and *Holocaust* (1978).

In the early 1980s, rogue television stations broadcasting from Monte
Carlo or islands off Britain further weakened European governments'
control of television. These pirates showed American and other syndicated programs, drawing viewers from public systems. When cable and
satellite TV joined the fray, the national systems partially capitulated in
order to retain some control over their home markets. France privatized
the first channel (TF1) in 1987 and then created a regional network
(France3), a cable system (Canal+), and a private network, La Cinq. By
1990 the number of worldwide TV broadcasters had so expanded that the
demand for programming was voracious. Televisions flickered in living
rooms from Texas to Turkey, while in Japanese and Mexican kitchens the
tube was on from dawn until midnight. Viewers also began to tune in Bol-

lywood musicals, Egyptian soap operas, Mexican *telenovelas*, and English League soccer. American exports were increasingly limited to the drama syndication, but competition appeared even there.

THE END OF U.S. DOMINANCE

By 1995 the introduction of high-quality, inexpensive videotape cameras and editing equipment as well as satellite or UHF broadcasting systems meant that a primitive TV station could be launched for a million dollars. Most African nations set up their own broadcasting systems and produced some of their own programming. The *Economist* noted that wherever "new stations establish themselves they tend to drop generic American products in favor of local productions: audiences still prefer homegrown fare if given the choice. In every European country in 1997, the most popular television programme was a local production."[35] The most popular drama in Jamaica was soon the domestically produced *Claffy*, in Zimbabwe *The Mukadota Family*, and in Nigeria *Mirror in the Sun*. Syria has become a major exporter of television dramas to the Arabic world, rolling out a dozen series that debut during Ramadan each year. War-ravaged Cambodia had three domestic channels and was producing a popular soap opera in Khmer when I visited in 2001. Even the number three station in Phnom Penh, CTV9, broadcast 47 percent local programming by 2002.[36]

To gauge the popularity of American television programs in Japan, in 2000 I assigned fifty seniors in my courses at a Japanese university to survey their families' television-watching habits in seven genres: variety, music, cooking, drama, news, travel, and film. Surprisingly, there were no imports at all in the top five positions of any category.[37] This was not a scientific survey, but it indicates the dominance of local tastes.

Recent trends have further undermined U.S. exports, chief among them "reality television." This was itself an import to the United States. Frederick Wasser writes that executive Jim Platt observed local programs in Australia, the Netherlands, and the United Kingdom that featured real-life police pursuit. This led to *America's Most Wanted*, a Fox hit of 1987. Fox created a marketing conundrum for U.S. producers. They had to respond in their home market, but reality programming was too culturally specific to be highly exportable. Soon France, Germany, and Spain had their own "survivor," "star search," and "roommate" programs, which trounced U.S. imports in ratings. A roommate show called "Big Brother Africa" now draws 30 million viewers from Kenya to South Africa to Malawi, where parliament has banned it for being too racy.

By 2002 Latin America used as much Mexican and Spanish programming as American. Asian and African stations mixed British and French shows with U.S. imports. Indian, Egyptian, and Mexican soap operas undercut the price of U.S. syndications and exploited a growing diasporic language market. U.S. invasions of Afghanistan and Iraq sapped the export market for U.S. news programming. By 2002 there were Spanish-language shows in all of the top twenty-five U.S. markets, and in California there were Mandarin, Cantonese, Korean, Japanese, Yiddish, and Russian channels. Mexico and Brazil had become the world's leading exporters of drama programs.

Meanwhile Fox, Time Warner, and other big program providers had moved on to what they believed was the next big thing, cable. They turned to China, the largest cable TV market in the world, with 98 million subscribers (only 35% of households). They vied with Japan's NHK and others to purchase a foothold, but by 2002 the rate of cable growth in China had cooled to 3 percent per year. Time Warner found it had overpaid for cable licenses that restricted it to diplomatic enclaves in the north and to the southern development zones near Hong Kong, where Cantonese is spoken. But only 50 million Chinese speak Cantonese, while 874 million speak Mandarin. Disney also found its future blocked by a 2000 Chinese law requiring that 60 percent of all cartoons on the air be made in China, and a 2005 proposal to ban any foreign animation between 4 and 9 p.m. U.S. producers must battle it out with Japanese, Philippine, and Taiwanese producers for the 40 percent market.

The promise of cable overseas turned out to be disappointing. Even in Japan only 26 percent of households had cable by 2002. In Germany, the figure was 50 percent—and that was the top rate for Europe. As Eric Pfanner wrote, "No amount of money or will could overcome the fact that Europe remains at least 15 distinct markets, with separate languages, tastes, business practices and regulations"[38] This is true worldwide: when stations broadcast seven days a week, twenty-four hours a day, language differences restrict audience size. In China, aggressive local entrepreneurs using smaller stations and satellite had snatched audience from big foreign companies by 2002. "For all the alarmist talk during the 1980s about wall-to-wall *Dallas* and *Dynasty*," David Puttnam writes, "the threat of all-U.S. TV failed to materialize and the popularity of American programming has, if anything, declined over the last decade."[39] By 2004 cable was everywhere losing ground to satellite television, which had lower land costs.

LOCAL PROGRAMMING

Seen a *telenovela* lately? Mexican entrepreneurs Emilio Azcárraga and Romulo O'Farrill are largely responsible for making these Spanish language soap operas into a worldwide genre. Televisa, their Mexico City company, produces 90 percent of the *telenovelas* shown in the United States and owns 11 percent of Univision, the major Spanish network in the United States. Its programming reaches every major American market, and the purchasing power of its viewers—$206 billion in 1995—was greater than that of the whole Mexican market ($166 billion). In fact, *telenovelas* have more viewers in the United States than at home. And in the critical eighteen- to thirty-four-year-old market, Univision beat at least one of the Big Four (ABC, CBS, NBC, Fox) every week in the 2005 season—and finished first over twenty evenings. NBC got so desperate that it hired Genesis Rodriguez from Mexico's number two Telemundo to star in a prime-time series. The two Mexican networks export to all nations of South and Central America, as well as Spain, Portugal, India, central Europe, Asia (even Japan is a market), and several Middle Eastern nations. There are Malaysian webzines for *telenovelas*.

Televisa was the largest producer of syndicated export programming in the world by 1990. It also bought *telenovelas* produced in other nations. Each *telenovela* "season" comprises 180 half-hour or 90 one-hour episodes (in series like *Corazon Salvaje* or *Dos Mujeres y un Camino*) and sells for about $200,000, according to the Paris newspaper *Libération*. With a cost of $3 million for 16 one-hour episodes of *Beverly Hills 90210*, program directors love *telenovelas*. The length of each season's syndication and its bargain price have helped *telenovelas* to displace U.S. drama exports in overseas markets.

There are now *telenovela* industries in Argentina, Brazil, Venezuela, and even Cuba. "Argentine soap operas," wrote the *New York Times*'s Moscow correspondent in 2001, "dominate daytime television here."[40] Peru's *Simplemente Maria*, a Cinderella story, began showing in distant Kazakhstan in 1994. By 2004 Brazil's Globo corporation was as big an exporter as Televisa, selling to 130 countries.

If we visit Televisa's studios in the Mexico City suburb of San Angel, we can witness something like the heyday of Hollywood. An army of 6,500 directors, actors, scenarists, technicians, makeup artists, and gofers work in eleven studios, the largest production facility south of the U.S. border. Episodes are designed for easy dubbing, and the filming pace is fast. Three cameras roll, controlled by a switchboard, as each scene is acted. There is a maximum of three takes. Actors don't memorize scripts, but read them from a teleprompter. As an actress named Maribel told *Libération*, "I did a little film work, but the Mexican movies got really bad . . . only TV could make me as popular as fast. For us Latin American actors, Televisa, it is our Hollywood."

Emilio Azcárraga Vidaurrea founded Televisa in 1952, and in 1955 started the first Mexican television network, Telesistema Mexicano. During the 1960s he bought up other Mexican TV and radio stations and invested in a variety of North American media companies, including cable. Before his death in 1972, *Fortune* ranked him as the richest man in Central America. His son, Emilio Azcárraga Milmo, made the company even larger, and his grandson Emilio Azcárraga Jean is now expanding it again. In 1998 *Forbes* magazine rated the Azcárraga fortune at $3.5 billion, and since then it has grown considerably.

Televisa broadcasts on four owned and operated nationwide channels. The most prestigious is Channel 2, which brings home half the company's advertising revenues. It features a variety of *Good Morning America*–type shows from 7 a.m. until 4:30 p.m., when it shifts into almost six hours of *telenovelas,* then closes with the news and a movie. Under Mexican law, up to 18 percent of broadcast time may be commercials, and Channel 2 exploits this largesse. Televisa's Channel 4 features cooking, fashion, exercise, and "how to" shows from 6 a.m. until 2 p.m., when it shifts to old American syndications (*The Untouchables, Bonanza, Tarzan*) and major league baseball. However, Channel 5 is the second most profitable channel, showing cartoons from 7 until 9 a.m., followed by Televisa sitcoms until noon. After lunch there are nine more numbing hours of cartoons, all imported oldies (*Captain Planet, Alvin and the Chipmunks, Power Rangers,* etc.). Many appear to be "American," but in fact they are Sony products created in the Philippines. At 9 p.m. Channel 5 concludes with a movie. This powerful channel, aimed at children and homemakers, covers 70 percent of Mexico and attracts a 22 percent national audience share. Televisa's fourth station is Channel 9, received in most of the country via two hundred relay stations; it broadcasts educational programming from 8 a.m. until 2 p.m., when it also launches a series of *telenovelas.*

Because Mexico is a neighbor of the United States, and its television system is modeled on ours, it offers a test case: how much of its programming is United States–generated? On a typical day (Monday, May 22, 1995), Televisa started its four channels at 6, 7, or 8 a.m. and signed off between midnight and 2 a.m. It broadcast a combined 73 hours of programming. Of this, "American" content was 27.5 hours, about one-third.

Of those 73 hours, 6 were movies (*My Science Project, Out of Bounds, Power*). The largest single category of imported U.S. programming was 12 hours of cartoons or children's shows run on Channel 5. This channel also ran two of the U.S.-made movies in the evening. The second largest user was Channel 4, which played *Tarzan, Bonanza, Hulk, The Pioneers,* and *The Untouchables* between 2 and 8 p.m. At 4 p.m. it showed an American film and at 9 p.m. part of a U.S. baseball game. Channel 2, the most profitable, showed only one U.S.-made program, *Wheel of Fortune*. The rest of its programming was Mexican. Channel 9, the educational channel, used no American shows. In sum, "American" programming was in three areas: cartoon shows; old syndications of cowboy, police, and adventure shows; and mediocre films.

Televisa accounts for four of the nine free stations in Mexico City. Of the others, Channel 7 (XHIMT) broadcasts a mix of home shopping and paid commercial shows, with music videos and major sports. Channel 11 (IPN) had only two attractions, a two-hour children's show called *Ventana de colores,* a sort of Mexican *Sesame Street* that was repeated during the day, and old films, such as Clint Eastwood's *Escape from Alcatraz* (1979) and Orson Welles's *Voyage of the Damned* (1976). Channel 13 (XHDF) was Channel 5's rival in cartoons, and it had the popular *X-Men* and *Nintendomania,* both Japanese syndications. In late afternoon it switched to Spanish and Mexican movies and variety shows. Channel 22 was a PBS-type channel that didn't come on until 5 p.m., showing classical music, jazz, and documentaries. Channel 40 (CNI) appeared at 6 p.m. with news, lottery results, and nature shows. Together these five channels broadcast sixty-five hours on Saturday, May 20, 1995, of which fourteen hours were U.S. made, less than a quarter of their airtime. Three old movies constituted six hours, cartoons four hours, the NBA playoffs two hours, and music videos two hours (the latter were 50% American).

This is the television universe for most Mexico City residents, though efforts were being made to sell cable to them too. The cable package looked like what we would receive in an American hotel—ESPN, CNN, Fox, MTV, USA, HBO, Discovery, NBC—except that most English channels were subtitled or dubbed. Multivision had the lead; it offered twenty-two channels, of which fourteen were in Spanish. The other cable purveyor was Televisa-owned Cablevision, which offered twenty-four channels, including Deutschevele, ABC, CBS, Cinemax, the Cartoon Channel, and CABLESPN. Ten of its channels were in English. Both systems were set up on the basics, premiums, or pay-per-view system familiar to Americans. U.S.-made films dominated the pay-per-view category,

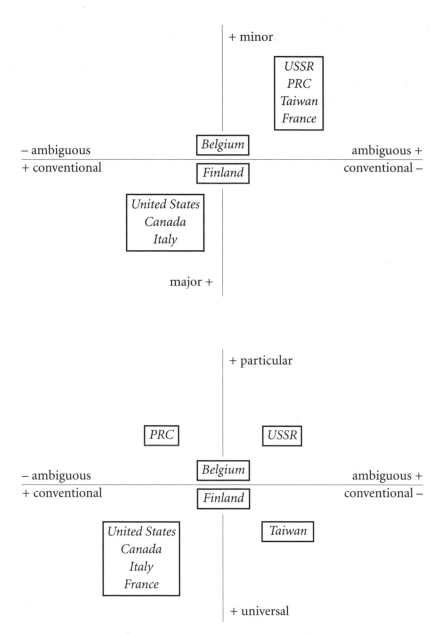

FIGURE 1. Divina Frau-Meigs's study of television narratives shows that even when cultures share a language, they may favor different themes or complexity. The upper figure depicts comparative narrative strategies; the lower figure, cross-cultural narrative options.

SOURCE: Divina Frau-Meigs, "The Cultural Impact of American Television Fiction in Europe: Transfer of Imaginary Worlds or Cultural Compatibility?" a paper presented at the European American Studies Association, Warsaw, Poland, March 21–24, 1996.

taking sixteen of the twenty-four hours available on Multivision and ten of eighteen hours on Cablevision. On two of the Cablevision channels, however, the same film repeated over and over all day. People to whom I spoke said that only rich Mexicans subscribed to pay-per-view.

The "American" programs that most Mexico City residents saw, therefore, fell into three categories. First, there were cartoons and kids' shows dubbed into Spanish. This programming bore an American logo and is reputed by Marxist critics to be especially pernicious, but it is mostly created outside the United States. Second, there were second-rank U.S. films, especially action, police, and horror. Like cartoons, these are difficult for other producers or distributors to generate in sufficient quantity. Both genres are easily dubbed. Given the need to fill the airtime of Mexico City's nine free and forty-six cable channels, this dependence is not likely to change. Third, there are older U.S. action shows in last-gasp syndication, such as *Magnum* and *McGyver*. Other countries, such as France and Germany, produce comparable syndications, but they are not as cheap as the American product, which at the end of its life can be offered very cheaply indeed. Historian of technology David Nye has estimated that it would cost foreign countries ten times as much to produce these series as it does to buy them.[41]

CULTURAL SPECIFICITY IN TELEVISION DRAMA

Recent scholarship contends that the values in television drama do not simply "transfer" to foreign cultures. Rather there has to be a narrative fit between program and viewer. The eminent French media scholar Divina Frau-Meigs, who has written four books on U.S. film and television in Europe, analyzed a large sample of U.S. television shown in Belgium, France, Finland, Italy, the USSR, the People's Republic of China, and Taiwan. Her methodology, based in communication theory and narratology, assessed the degree to which local programming veered toward the "American model" when it was present in the market. The latter was defined as having "fewer themes per program, strongly emphasized" with "little ambiguity" in character or plot development. An illustration would be an American police show. As Frau-Meigs writes, "Only two major styles emerge out of a combination that could provide several alternatives: either a multiplicity of themes with much ambiguity or a few themes with little ambiguity"[42] (see fig. 1).

As Frau-Meigs shows, the United States, Canada, and Italy tended strongly to few themes in plot and little ambiguity in character. But France, Russia, Taiwan, and China favored many themes and greater ambiguity.

No wonder that the old U.S. detective show *Columbo* is still playing in some of them. Finland and Belgium occupied the middle ground. When the programs were examined for the number of narrative complications, an index by which communication scholars judge a culture's relation to ambiguity, a different pattern emerges. France, Italy, the United States, and Canada were grouped together in the "conventional, universal" quadrant, whereas Russia, China, and Taiwan diverged dramatically. Russia stood alone in combining a great number of themes with character particularity and narrative ambiguity. Most surprisingly, Chinese and Taiwanese audiences were radically different, the latter preferring generalized characters in ambiguous plots, whereas the former wanted particularized characters in conventional plots. Frau-Meigs concluded that "contrary to intuition, geographical closeness as well as cultural and linguistic ties are not as influential as they might seem. The most striking example is that of Taiwan, which should be closer to the People's Republic of China, but is in fact closer to Europe and the USA. This phenomenon is also valid within the European block: Belgium doesn't align itself on France, its cultural and linguistic neighbor." Program exporters have tried to foist off series that were cheap, without regard for narrative style, on developing nations. But the exporter's cultural values were often too jarring. "The American programs," notes Frau-Meigs, "introduce the diversity of Otherness into local representation, but it is an Otherness that is not totally assimilated. If there is indeed contact, there is no proof of successful transfer."[43]

Rather than imposing a Western "fantasy world" on viewers, exported programming seems to take hold in niches of narrative compatibility. In Italy, where male television viewers are obsessed with showgirls called *velinas*, *Baywatch* is as popular as it was in the United States. But in Poland, to which Silvio Berlusconi donated his Raiuno channel after independence, Italy's *velina* television, with its buxom underdressed showgirls, never caught on, especially with devout Catholics. As for American programming, in China and Russia it is only 10 percent of content, whereas in Taiwan it is as much as 50 percent. Shows that were not blockbusters in the United States sometimes rise to fame overseas. *Columbo* is a case in point, but *Seinfeld* has never been as popular, probably because its comedy is culturally specific. And African American comedies like *The Cosby Show* hardly export at all. Indeed, foreigners' taste for multi-themed, complex-charactered, and narratively ambiguous programming has worked against the sale overseas of the typical American television drama (there's some evidence that *American* tastes are moving toward the

French-Russian pole). And at the simpler end of the spectrum (soap operas), American exports are undercut by *telenovelas*, which take the extended family as the basic social unit and focus on a romantic couple, a formula more appealing in the developing world.

CARTOONS

The influential argument about the impact of Disney products made by Ariel Dorfman and Armand Mattelart (*How to Read Donald Duck,* 1971) seems to have been undercut by history. They argued that the basic narrative of Disney cartoons comes from the United States and reflects an imperialist world view. First published in Chile in 1971 during the Allende years, their book shares the revolutionary tone of the period. It is an attempted "unmasking" of children's fantasy literature to reveal "capitalist, bourgeois" values, in much the same way that Marx analyzed Eugene Sue's popular novels. Subsuming a good deal of Freudianism, Dorfman and Mattelart postulate "childhood" as a "utopia" colonized by Disney, with parents, labor, sexuality, and procreation eliminated. In this view, the ducks are all coequals sent by Uncle Scrooge, who is a puppet of unseen capitalist forces, to exploit Unsteadystan, Inca-Blinca, and Aztecland. Their adventures teach them the proper colonialist attitudes and techniques. As readings, Dorfman and Mattelart's are not implausible, and many readers will find the portrayals of Third World peoples shockingly crude and the narratives distressingly nationalistic and ethnocentric. But more distressing still is Dorfman and Mattelart's conception of the foreign reader: "The housewife in the slums is incited to buy the latest refrigerator or washing machine; the impoverished industrial worker lives bombarded with the images of the Fiat 125. . . . Underdeveloped peoples take the comics at second hand, as instruction in the way they are supposed to live and relate to the foreign power centre."[44]

In the Marxian view, readers in "underdeveloped nations" are robots, believing whatever they read to be gospel. Fortunately, other critics have corrected this error. As John Tomlinson writes in *Cultural Imperialism,* "It would be absurd to assume that people in any present-day culture do not have developed attitudes to such a central aspect of their lives (as the relationship between wealth and happiness) quite independent of any [such] representations."[45] In fact, as far back as 1989 M. Barker argued convincingly that three other interpretations of Scrooge McDuck were just as plausible as Dorfman and Mattelart's: in one Scrooge McDuck is a satire on the mania for money getting; a second extends this to a closet critique of capitalism, with Scrooge as "a biting parody of the bourgeois entrepreneur"; and a third reading finds a larger theme about "the ways in which human beings deceive and destroy themselves."[46]

Equally important, television cartoons have not been produced in the United States since the late 1970s, because of the high cost of hand-inking the cells. Despite the increased use of computers and other short cuts, cartoon production is still labor intensive. Low-cost cartooning colonies developed in Seoul, Manila, Barcelona, Taipei, and Bogotá in the 1980s. By 2000 about 90 percent of all "American" cartoons were created in Asia, with the Philippines having recently supplanted South Korea as the capital. "Over the past 10 years, four major animation companies employing more than 1,700 people and several smaller studios have opened in Manila alone," writes Michael Switow. "Familiarity with U.S. culture gives Filipinos an advantage over other Asian competitors in the animation field."[47]

Cartoons are now developed in the following manner. Major studios, such as Sony, Disney, Marvel, Warner, and Hanna-Barbera, send storyboards and voice tracks to Manila. Filipino artists and technicians draw, paint, and film about twenty thousand sketches, mainly by hand, for every thirty-minute episode. The material is then sent back to the United States or Europe or Japan, where the sound effects and music are added. The Disney shows *Timon and Pumbaa,* a *Lion King* spin-off; *Duck Daze* featuring Donald Duck; and *Aladdin,* a spin-off of the movie, are among the cartoon shows produced in the Philippines. Drawing the cells of a thirty-minute cartoon would cost $500,000 in the United States, and $200,000 in South Korea, but it costs only $165,000 in Manila. "Fred Flintstone is not from Bedrock," Switow writes, but from Manila. "So too are Tom and Jerry, the Biker Mice from Mars, Aladdin and Donald Duck."[48] Philippine studios, inevitably, have begun to create their own cartoons: Fil-Cartoons was producing two series for the Cartoon Network by 1996.

The editing and printing of the Donald Duck comic books is even further outsourced. David Kunzle's updated introduction to the 1991 edition of Dorfman and Mattelart's *Donald Duck* reveals the multinational nature of these texts. There are "at least four different Spanish-language editions of the Disney comic," he writes.[49] The Chilean edition serves Peru, Paraguay, and Argentina, but printed only around 800,000 copies a month at its peak in 1970. Of the 4,400 pages of material, one-third was created by Disney studios, one-quarter came from Italy, one-third from a U.S. franchisee (Western Publishing Company), and the rest from Brazil and Denmark. The Mexican series, only 2,200 pages total, received all of its material from the United States. The Brazilian edition, Kunzle notes, is fairly dependent upon Italy for material. Italy supplies 1,000 of the 5,000 pages in the five Brazilian titles. Brazil itself generates another 1,000 pages. Another edition originates in Colombia, serving northern South America. Italy originates over half of the pages in Italian Disney comics, and France half of its own pages. In the final analysis, as Kunzle shows, Donald Duck is only a logo, a franchise.

Like other good franchises, it seems that Donald is adaptive, a different duck outside the United States. Local editors do what they want to sell the product to local audiences, like "fusion cuisine" chefs in the United States. They alter scenes, invent new characters, and change the dialogue. In some cases, what remains of "Disney" is the physical appearance of the characters and the name. In some countries, of course, the product is substantially the same. But clearly "Donald Duck" is *adapted* worldwide to fit local cultures. He becomes "Pato Donald" or "Unkle Donald." As such, he is no longer an ideological vehicle of Disney values, or even of consumerism, but of a hybrid of some of these and the local cultural values.

The McDonald's Brouhaha

Is the world suffering from "McDonaldization"? Critics offer no statistics, no studies, and few facts to back their generalizations. Write George Ritzer and Elizabeth L. Malone, "The most notable and more directly visible cultural impact is the way McDonald's is altering the manner in which much of the rest of the world eats. What and how people eat is a crucial component of almost all, if not all, cultures, but with the spread of the principles of McDonaldization virtually *everyone* in McDonaldized society is devouring French fries (and virtually every other kind of food) and doing so quickly, often on the run."[50] I am not arguing for junk food here, but I do think overseas fast-food emporia deserve objective study and analysis. The known facts would suggest that McDonald's is at least as representative of *modernity* as of Americanism, and that the former rather than the latter is responsible for changes in traditional eating habits.

McDonald's opened its first foreign franchises in the 1960s, and it was soon followed by its competitors. The perception of a rise in "McDonaldization" owes much to a concurrent increase in American tourism, beginning with the 1960s backpackers. Even today, if we venture somewhere that Westerners don't go—Greenland, Nigeria, New Guinea—we find few McDonald's. But "McDonaldization" has become the rallying cry of a wide variety of modernity's foes. Food franchises are undeniably a feature of modernity and globalization, but how "American" are they?

It takes only a minute's reflection to realize that *every society has always had a form of fast food*. On Bali, one pot of food was traditionally cooked in the morning, and everyone snacked from it all day: a fast meal was rice and meat carried away on a palm leaf. In France open shop fronts have sold *crêpes,* sandwiches, and *croque-messieurs* at least since the 1940s. Today *kebab* stands dot every French town, offering carry-away meals. Germans and Austrians can buy chestnuts and *kartoffeln* from street-corner vendors if they need to eat on the run. Mexico is overrun with street-corner taco and burrito vendors. In Japan most people eat boxed meals of rice, vegetables, and protein called *bento* for lunch. *Okonomiaki* and *takoyaki* carts still punctuate some street corners, while *soba, udon,* and *ramen*—once sold by itinerant street vendors—are everywhere available in storefronts. *Sushi* was originally fast food, sold from pushcarts, and there are now robots that turn out 1,200 pieces an hour. Fish and chips predated McDonald's in Britain by a century. North American Indians had *pemmican.*

The idea that Americans invented fast food would be hilariously ethnocentric if it were not so widely believed outside the United States. Equally uncritical is the notion that United States fast food *causes* foreigners to eat faster than they used to. The Japanese have always eaten lunch quickly, and Mexicans are no laggards. Many Europeans simply skip lunch. Some cultures eat faster than Americans do, and some eat more slowy. The pace is governed by factors other than proximity to the golden arches.

Yet there are professors asserting that McDonald's caused the demise of sit-down dining in Japan. A strange charge, since for more than a hundred years Japanese has had words for eating while standing (*tachikui*) and drinking while standing (*tachinomi*). In fact, the oldest bars in Japan are the *tachinomiya* (place to drink standing up), and all railroad stations have had standup *ramen* shops since at least World War II. Historic accounts, drawings, and photos show that the Japanese ate while standing in the street in the 1880s.

Everyone, of course, feels a right to weigh in on the fast-food debate—after all, they eat it!—but almost no one goes out and studies eating. And critics in the United States do not recognize the increase in foreign foods that they themselves eat, tastes that began to return with those backpackers in the 1960s: *tom yum* sauce from Thailand, *fleur du sel* from the French Camargue, Jamaican rum, and Italian tomatoes. Augmented by increased incomes and personal mobility, American tastes led the way to a *modern* preference for a more diverse food palate, which has since spread to the rest of the world. But this modernization has not been accompanied by the spread of "American" flavors. As for McDonald's, it has succeeded abroad largely for the same reasons that it has succeeded at home.

McDonald's tried a "store" (to use company parlance) in Holland in 1960, which failed, and then another in Canada in 1967, which succeeded. It opened its first Japanese store in 1970, and by 1980 it had four hundred there, where they are known as "Makudo." By year 2000 there were four hundred in France. In Italy, home of the "slow food" movement, 500,000 Italians a day were eating at three hundred McDonald's by 2003. That is actually not very many.

McDonald's *is* a success, but is it worth apoplexy? Most overseas McDonald's (and other fast-food franchises) still cluster around the most touristed parts of Tokyo, Paris, London, and other major cities because Americans and other foreigners, uncertain about local restaurants and familiar with McD's low prices and cleanliness, end up eating there. This added business not only perks up sales but gives the franchise an aura of authenticity. If we eat fast food at home, it is not surprising that we find it abroad: Japanese and Chinese and Frenchmen visiting the United States can find their fast food here too. But even in Japan, where it has nearly 1,500 outlets, McDonald's and its cohort have a smaller footprint than critics realize.

Most critics of McDonald's don't understand the basics of fast-food retailing. Here are two of them "deconstructing" McDonald's as it "infiltrates local culture":

> Beijing customers often linger for hours rather than eating quickly and leaving or taking their food with them as they depart the drive-through window, which undermines one of the principal dimensions of McDonaldization—efficiency. Perhaps the biggest difference, however, is that in Beijing McDonald's seeks to be more human by consciously presenting itself as a local company, as a place in which to "hang out" and celebrate important events and ceremonies (e.g. children's birthday parties). Rather than simply a place to get in and out of as quickly as possible, personal interaction is emphasized by employing five to ten female receptionists, who are referred to as "Aunt McDonalds" (similarly Ronald McDonald is known as "Uncle McDonald" in Taiwan), whose main tasks involve dealing with children and talking to parents.[51]

In reality, McDonald's has always, everywhere, welcomed customers to stay a while. That is part of its marketing plan. In small U.S. towns McDonald's is the location of business meetings, Rotary gatherings, after-the-game analyses, and even Gospel sings. Birthday parties with hostesses, special children's sections, and a willingness to let orderly teenagers hang out have been part of its home habit for decades. It sponsors Little League teams and donates to the cheerleaders' uniform fund. The description above of foreign McDonald's could be a franchise in Arlington, Texas, or on New York Avenue in Washington, D.C. Furthermore, drive-through windows, which these scholars think are "efficient," are, like parking lots, very expensive and inefficient for all fast-food sellers.

Here are some statistics from a profile by Stephen Drucker in the *New York Times* that help to put McDonald's in perspective.

- Only 7 percent of the U.S. population stops by daily—and this might be only for a cup of coffee.
- McDonald's accounted for only 15.2 percent of the fast food sold in the United States in 2002, down from a peak of 18 percent in the late 1980s.
- In 2003 the typical customer was a male, from midteens to early 30s, who ate there twice a week. He accounted for 75 percent of McDonald's U.S. business, and he expects—really!—to be served within one minute, though the company meets that expectation only about half the time. He is, obviously, a narrow slice of the U.S. population.
- Half of McDonald's dollars come from window sales, but auto traffic requires more land and parking lots and high-tech order systems. These are more labor intensive and costly than counter sales.
- In 2000 there were about 12,000 McDonald's in the United States, and 8,000 in eighty-nine other countries. There are 1,482 *Makudos* in Japan (by far the largest foreign presence), 430 in France, 63 in China, 2 in Bulgaria, and 2 in Andorra. These were owned mostly by foreign franchisees.
- As for Beijing, only 10 percent of the Chinese population can afford to buy a Big Mac, by McDonald's estimates. The company is there because it wants a foothold in the world's largest market.
- McDonald's spends about $1 billion a year on worldwide advertising. In the 1990s, most of that went for television ads shot by Leo Burnett and DDB Needham in an upbeat "Steven Spielberg style." This advertising is not localized, often produced abroad, and probably less effective than it should be. Its ineffectuality is a sore spot with McDonald's shareholders and franchisees.

• McDonald's customers are *very* picky and attempts to win them to new products fail more often than they succeed. The McLean Deluxe, introduced in 1991 because U.S. critics carped about fat, was a major failure. "People talk thin but they eat fat," says senior vice-president Richard Starrman.

Perhaps the biggest misconception is that McDonald's is the Trojan horse of American ideology. But golden arches out front do not translate to "American" inside, as even its critics note: "McDonald's adapts to each distinctive cultural context and, as a result, is so modified that it is ultimately impossible to distinguish the local from the foreign. Thus, in China McDonald's is seen as much a Chinese phenomenon as it is an American phenomenon. In Japan McDonald's is perceived by some as *Americana as constructed* by the Japanese."[52] Though not a critic, Thomas Friedman made the same point when he noted a Japanese child visiting the United States who was surprised to learn that there were McDonald's here. This is an aspect of what James L. Watson has termed the "transnationality" phenomenon, in which a company becomes a federation of semiautonomous enterprises.

But even transnationality has limits. McDonald's franchisees may add beer in Germany, salsa in Mexico, and soy flavors in Japan, but the essence of McDonald's is its process and logistics. It always has low prices, a clean dining room, efficient service, polite staff, good lighting, lots and lots of free seating, even for noncustomers, and free, clean bathrooms. This may seem obvious to Americans, but in much of the world this is revolutionary. In the rest of the world one simply does not enter a restaurant without buying something, much less use the toilets (for free!)—cleanliness in the latter would be iffy anyway. There are certainly no free seats for doing homework, as in the Nishinomiya Makudo. Even Ritzer and Malone concede that "in both Hong Kong and Taipei McDonald's virtually invented restaurant cleanliness and served as a catalyst for improving sanitary conditions at many other restaurants in the city."[53] The same is true across Europe, not to mention Mexico and South America.

Like a classic French restaurant, McDonald's allows customers to stay as long as they want. When I taught in Vienna in 1993–94, McDonald's had to raise the price of its coffee to the level of Viennese *Kaffeehäuser,* because the latter complained. It seems that elderly *omas,* who were hustled rudely out of the local Meinl coffee shops, discovered they could spend an afternoon at McD. In Mexico, Japan, France, Taiwan, and Poland, I've seen teenagers hang out at McDonald's after school for hours, doing homework or talking. Travelers stop in to read undisturbed, businessmen to call home or telecompute via cell phones. All this in a "clean, well-lighted place"—Hemingway's old man knew how rare they were. It is a value of modernity that is appreciated worldwide.

On one hand McDonald's is accused of standardizing international taste. But aren't clean, free bathrooms a good standard? On the other, when critics discover that McDonald's alters its menu to suit local tastes, it is accused of an insidious capitalist plot: McDonald's "impact is far greater it if infiltrates a local culture and becomes a part of it than if it remains perceived as an American phenomenon superimposed on a local setting," write Ritzer and Malone.[54]

Few critics realize that the chain's great successes, from the Egg McMuffin to the Big Mac, sprang from *local* franchisees, who are encouraged to experiment. And foreign franchisees are not getting hamburger from America; they have to find local suppliers as soon as possible, buying their meat and potatoes, their milk and buns in the area. McDonald's of Austria even taught farmers in Poland and Slovakia how to raise the low-water-content potatoes used in McDonald's fries. McDonald's franchises in Austria buy 90 percent of their ingredients in central Europe.

But there are a dozen other American fast-food chains abroad. As Thomas Frank wrote in the *New York Times:*

> Even more adaptive in terms of food are the smaller American food franchisers (Big Boy, Dairy Queen, Schlotzsky's Delicatessen, and Chesapeake Bagel) that have followed McDonald's and the other American giants overseas. In 1990, alone, these mini-chains opened 800 new restaurants overseas and as of that year there were more than 12,000 of them in existence around the world. However such mini-chains are far weaker than McDonald's and therefore must be even more responsive to local culture. Thus, Big Boy sells things like "country-style fried rice and pork omelet" and has added sugar and chili powder to make its burgers more palatable to its Thai customers. Because it caters to many European tourists, it has added Germanic foods like spätzle to its menu. Said the head franchiser for Big Boy in Thailand: "We thought we were bringing American food to the masses. . . . But now we're bringing Thai and European food to the tourists.[55]

Local entrepreneurs have caught on quickly, blending more local cuisine with fast food's speed, modernity, and service. In China there are three imitators of KFC alone: Ronghua Chicken, Xiangfei Roast Chicken, and

Beijing Fast Food Company. The founders of the latter used to work for McDonald's and KFC.

If we return to the Nishinomiya train station, we can see a Moos Burger (1,500 outlets in Japan), which serves a sloppy-joe concoction, and a Yoshinoya, which serves traditional Japanese food fast (more than 2,000 outlets—more than McDonald's!). In three minutes Yoshinoya serves up a salmon filet, vegetables, rice, pickles, and tea for $4.50. In 1979 it opened its first U.S. shop, and there were sixty-two in Los Angeles County by 2003, with plans for a thousand nationwide.

In Russia there is Russkoye Bistro, which has more than a hundred outlets and serves 35,000 to 40,000 customers per day. "If McDonald's had not come to our country," says Russkoye's deputy director, "Then we probably wouldn't be here. We need to create fast food here that fits our lifestyle and traditions. . . . We see McDonald's like an older brother. . . . We have a lot to learn from them."[56] When I visited impoverished Latvia just after the Iron Curtain lifted, there was one McDonald's with three local imitators. The most successful was called "Little Johnny's," run by former employees of McDonald's, and it was doing a better business than its older brother.

Let's take a wider view. There are far more restaurants in the world now than there were twenty years ago, and there are still many more traditional restaurants than there are fast-food restaurants. One need only wander the back streets of Paris's thirteenth arrondissemont, Tokyo's Asakusa, Marseilles' harbor front, or Vienna's ninth district to realize that outside the tourist precincts, fast food is not that common. And there are still millions of locally owned bars and restaurants. Critics seldom consider the size of the restaurant universe. In Paris there are more Chinese restaurants (1,500 according to a recent article) than fast-food franchises of all types. The mayor of New York likes to boast of his city's seven thousand restaurants, only about a thousand of which are fast-food franchises. In other countries the percentage is even smaller. The twenty square block area of Osaka's Minami-ku is reputed to house five thousand restaurants—and none of them are Western fast-food franchises. Tokyo has more than ten thousand local restaurants. Guangzhou (Canton), China, has seven thousand local restaurants.

In Avignon, France, where I have lived twice, there is only one McDonald's downtown. It draws tourists, French teenagers, mothers with children, and local business people for breakfast, which is hard to find if you want more than a croissant. If you need a bathroom or a quick cup of coffee, McDonald's is the place. But there are two hundred other restaurants and bars in central Avignon, most of them locally owned. After 7 p.m. they do a huge business, while McDonald's is closed. Fast food in Avignon is hardly new. *Plats à emporter* is what the *kabob* merchant in Place Pie sells. On the Rue Carreterie the charcuteries have been selling *patés* and takeout dishes, and the *boulangeries* have been selling *baguettes* and sandwiches for a very, very long time.

Fast-food emporia have another upside. They run management training programs overseas that give local managers the equivalent of an MBA. This knowledge about how to run a business, as the Russians testify, is invaluable. McDonald's, KFC, and Wendy's all sponsor such programs. KFC runs a "university" in Kuala Lumpur, Malaysia, where promising young managers from its South Asian stores learn management skills. The only other institutions there that teach these skills are universities for the rich. Like the cleanliness, however, this training is more modernity than Americanization.

There are other values associated with fast food that we might wish to see more of abroad. By now it is a cliché to speak of the way in which the industry socializes youth to the workplace; provides opportunities for minorities, the handicapped, and older citizens; or sponsors local charities and fund raisers. But it is worth noting that the family-owned "greasy spoon" of yore did none of these. It did not promote sexual equality in the workplace, nondiscrimination, handicapped access, or corporate charity. These ideals, where practiced overseas by McDonald's and its imitators, cynically or not, are new to most developing nations and some European ones. Fairness, compassion, and meritocracy are still a tough sell abroad, however, and are widely resisted, ignored, or resented.

Critics also fail to understand that most fast-food franchises are locally owned. Most of the profit, power, and experience stay abroad. McDonald's selects locations, based on human and vehicle traffic and other considerations. It trains franchisees extensively, then offers them locations. McDonald's owns the land, so there is no chance for franchisee self-dealing in real estate. Its real-estate acumen offers McDonald's as much opportunity for profit as its cut of the franchisee's sales; indeed, to Russia's risky realty market McDonald's has brought traffic analysis and other sophisticated tools now copied by locals.[57] The franchisee must install exactly the shop that McDonald's stipulates and go to work in it full-time. Hands-on management is the norm. These are radically different practices from those that prevail in most of the underdeveloped world. The result is that some urban franchises make $2,000 an hour during peak lunch periods. As for food quality, McDonald's operates an extensive customer feedback and complaint system. There is less than one complaint about food per store per year worldwide, and only three about service. The facilities themselves receive a minuscule number of complaints, according to Drucker.

Saddest of all, critics don't realize that McDonald's is not the world's largest fast-food company. That honor goes to the Compass Group of Great Britain, which owns Burger King, Sbarro, and a host of other brands. Compass serves more airports, company lunchrooms, and school cafeterias than any other company, but it has no single, recognizable logo. McDonald's is second, followed closely by the French firm Sodexho. It seems that while some Frenchmen are criticizing the golden arches, other Frenchmen have a $1 billion a year contract to serve food fast to the U.S. Marine Corps, the UN forces in Kosovo, and American forces in Iraq. In fact Sodexho employs 110,000 Americans; it's a global power, even a colonial one. McDonald's employs only 35,000 French. All facts considered, the McDonald's brouhaha is more about image than substance.

What about the Internet?

Is the Internet mostly Anglophone and available only to the affluent? Is the Internet really enriching? Critics contend that, to their detriment and our advantage, underdeveloped nations don't have sufficient access to the Internet. But a look at the historic development of the Internet and current-use statistics suggest that the rest of the world is right behind the United States. Not only does a nation not have to be affluent anymore to have Internet access, but the narrowness of Americans' Internet use is striking, calling into question its "enrichment" power. Pornography often constitutes as much as 40 percent of North American Internet traffic, on-line file-sharing of music and video another 30 percent, according to some Nielsen/NetRatings. And on-line video gaming has risen to 10 percent, as broadband connections have made that pastime more feasible. These three areas may account for 80 percent of U.S. Internet activity. At some U.S. universities 50 percent of bandwidth is sometimes devoted to video- and song-swapping programs. In July 2003 the most heavily trafficked nonportal, nonnews Internet site in North America was eBay, the on-line flea market.

"We always think of the Internet as being very diverse, democratic— that everyone goes to hundreds of sites every week," says Mark Mooradian, senior analyst at Jupiter Media Matrix, a company that measures web traffic: "In truth, that's less and less the case." North American use has grown particularly narrow. In 1999 more than 60 percent of U.S. users visited twenty sites a month, but by 2000 the same percentage visited only ten sites a month. In July 2001, Jupiter reports that a few top-ranked websites dominated the most popular genres. In news—32.1 percent of surfers went to MSNBC, and the top three news sites—MSNBC, CNN, and the *New York Times*—pulled in 72 percent of the news traffic. To conduct a search, 30 percent of Americans went to Google alone. For weather, 64 percent went to Weather.com. To find a map, 82.4 percent went to Map-Quest. In most categories, the top-ranked five sites accounted for more than 90 percent of all searches made by North Americans in 2003. "I guess I feel I've found most of the things of interest to me," writes a man interviewed by the *International Herald Tribune*. "Surfing jaunts tend to feel like bicycling around the block," he adds. "I'm also much more pointed in my Web use—I typically get some durn-fool notion in my head . . . and fire some queries into Google and click until either the subject seemed exhausted or I am." As Joseph Turow, professor at the University of Pennsylvania, told the same newspaper, "people are encouraged to drill down into their areas of concern to such a degree that they get closeted in their own reflections of themselves."[58]

Nor is the world's most wired nation the United States. The champion is South Korea, where 50 percent of households have Internet connections and a much higher percentage than in the United States, one in sixteen, were connected by broadband at the end of 2000. In the United States only one in forty-eight households was similarly connected. Visiting Seoul in 2000, I saw computers everywhere. I was lured into Kinko's to read my e-mail, then learned there were ten thousand cyber cafés in South Korea, many in back alleys and in small towns. What South Koreans *do* with their computers is different than what North Americans do. Seoul mayor Goh Kun ordered up a website called OPEN in 1998 that allows ordinary citizens to trace their applications for building permits, business permits, and alcohol permits through the government bureaucracy. Not only does the system tell on whose desk the application sits, but it requires denials to be explained in writing on line. Kofi Annan is such a big fan that the UN is translating the Seoul software into six languages.

Of course, a good bit of South Korea's high-speed access is devoted to something that critics find deplorable—video gaming. South Koreans are the world's most expert players of interactive on-line games such as FIFA2001 (soccer) and Starcraft (combat). FIFA2001 is popular throughout Europe and Asia, but in South Korea 4.5 million people log on and play daily. That is 10 percent of the population. "A waste," you might say, but it has led to an Internet infrastructure "as sophisticated as anything in the United States," writes Gregory Beals, and it has doubled, to twenty, the number of South Korean firms that write game ware. South Korea has professional video-game players who make $60,000 a year, and pro teams, such as Samsung's "Khan." Korea Telecom Freetel and KTB Network (the nation's largest venture capital firm) are betting that South Korea can be a major player in digital entertainment. In short, this deplorable activity has generated an industry.

Malaysia used its oil wealth from the 1990s to invest in computers. The result is not only a high-tech corridor around Kuala Lumpur but also computers in the smallest towns. Kota Bahru (population 8,000) sits on Malaysia's restive northern border with Thailand, the center of a Muslim renaissance. Yet in 2001 I found three cyber cafés there, two of them run by and catering to Muslim women. In these cafés the women keep in touch with out-of-town family and friends by e-mail. One of them told me, "This is faster and cheaper than writing letters." Her cost was eighty cents per hour. This cyber café also offered classes in Word, Excel, and Access, all in the Bahasa Malaysia language. There were also cyber cafés in the Chinatowns of Penang and Malacca, and in the distant Cameron Highlands. A few months later I logged on from an even more remote location, one of two cyber cafés in Mataran, on the impoverished island of Lombok in Indonesia. In none of these places were other Westerners among the users.

In Malaysia an ambitious NGO (nongovernmental organization) official named Gabriel Accasina has put together a mobile Internet bus program for rural areas. He has eight, twenty-position mobile computer labs that tour provincial schools on a fortnightly basis. "The bus program is typically built around an eight-hour course delivered in one-hour installments to 20 children at a time," writes Wayne Arnold.[59] "It starts with such basics as learning how to turn the computer on and use a mouse, then progresses to basic word processing, e-mail, Web browsing, even manipulating spreadsheets and designing simple Web pages." Muslim clerics opposed the "frivolity" of the program, so Accasina added before and after tests. They showed dramatic increases in the ability to use a computer, but also a significant drop in spelling errors and increased reading comprehension. The computer bus leaves a computer loaded with a mini-Internet and a stack of CDs at every school. At the remotest schools, which cannot be visited every two weeks, Accasina leaves up to ten computers. The UN's International Development Program, which is the cosponsor, has organized a similar initiative for Ghana.

Free enterprise may beat the UN to Ghana. With Ghanaian business-men as his silent partners, Mark Davies, a former dot-comer, opened BusyInternet there in September 2001. Housed in a former factory, his 14,000-square-foot facility in Accra offers low-cost public access on Pentium III computers with flat screens and satellite connections. It has training facilities, meeting rooms, and photocopy machines. Davies and his partners have more than $1 million invested in Ghana, and they plan to build next in Nigeria, Ivory Coast, and Uganda. This might seem like a risky venture, and it has its critics. But Davies says he was persuaded by statistics from NUA Internet Surveys, which estimated that there were twenty thousand Internet users in Ghana in 2000. "There are 240,000 telephone lines in Ghana for 19 million people," Davies told the *New York Times*. "It takes about seven dials to make some phone calls go through, just across town. We've put in our own link to the national electric grid, our own generator, our own satellite dish for bandwidth. Our philosophy is to say nobody really knows what's right for Ghana, and the technology is sort of culturally specific in terms of how it's implemented and how it works."[60]

Davies is not the first Internet provider in Ghana, just the fastest. The poor telephone system spurred this development. By 2001 there were already four other providers and more than one hundred cyber cafés. Although some of Ghana's Internet traffic still depends on land lines, an increasing part bypasses it. A deep-sea data cable from Africa to Europe is scheduled for completion, funded by forty international telecommunications companies, and Africa One, a private company, has plans to build a fiberoptic cable across the width of Africa. If all these plans gel, Ghana and other African nations may skip old-fashioned telephones.

In Russia, too, dated infrastructure has been an obstacle. There are unbelievable distances to span, so the Internet is basically available only in the East and Southeast. In a typical week of 2001 only 2.5 percent of Russians were on line. That figure grew to 8 percent, or 11 million users, by 2003. The future is unclear. Actually, to say that infrastructure is an "obstacle" in Russia is an understatement. When I traveled there in 1994, there was one antiquated phone line per communal apartment, and some-times only one per building. There were power surges and dips daily, and my telephone calls were often cut off. Outside of tourist hotels, which had installed their own phone systems, there were public phones that dated from the 1960s and required a one-ruble coin, which cost several hundred rubles to buy. While the spread of the Internet in the West is buttressed by logistical tools, such as credit cards, encryption, UPS, and FedEx, little help exists in Russia. The *New York Times* reports that "it is hard to find viable Internet projects run by capable and responsible entrepreneurs. Everyone in Runet [Russia] dreams of doing an I.P.O. as in the West, but they believe this to mean some rich sugar daddy comes along with a bag full of dollars and that they need not be accountable for this investment."[61]

What do Russians do on the Internet? They log on the entertainment or news sites, much like Americans. According to Russian analysts, current growth in demand is coming from more-remote provinces, which will be hard to service. Another obstacle is that even the best deal—$20 a month for unlimited hours in 2002 (down from $40 in 2000)—is still too expensive when the average Russian earns $50 to $100 a month.

HOW MUCH OF THE INTERNET'S CONTENT IS ENGLISH?

Sometime in 2003 English ceased to the language of the majority of Internet pages. Computer scientists saw this coming, but critics of globalization did not. Most computers in the United States are set up to display only the Latin alphabet; only a few even display Spanish correctly. Search engines such as Google and Dogpile, unbeknownst to most of us, come in various language flavors. The Spanish versions of most search engines prioritize Spanish sites, the English ones English sites, and so on.

There is a lot of information on the English content of the Internet, less about other languages, but experts agree that the non-English Internet is growing very quickly. One way to gauge growth is the registration of new domain names. Matthew Zook of the University of California at Berkeley follows this information and found that in 1998 about 49 percent of new domain names were registered from the United States. That rose to about 55 percent in 1999, but then it began to drop. By 2001 the United States accounted for only 40 percent of new domain names. Great Britain and Germany were second and third, each with about 10 percent. Canada, South Korea, and the Netherlands followed. The world's second largest economy, Japan, ranked a surprising ninth in new domain names. According to Zook, the total percentage of domains ending in .com, .org, .net, and .edu attributable to English-speaking nations declined from 74 percent in 1998 to 59 percent in 2001. The U.S. share of .edu domains—the websites of educational institutions—was 85 percent in 1998, but within four years fell to 72 percent.

There are other ways to get a grasp on the Internet. Zook also combines information from the CIA, Nielsen/NetRatings, and the Computer Industry Almanac to track the percentage of a nation's population "on line" with some kind of home Internet access. The United States, Australia, Canada, New Zealand, Great Britain, Japan, South Korea, Taiwan, Switzerland, Austria, and the Scandinavian countries have led the way, with more than 35 percent on line since 2001. Germany, France, Italy, Ireland, Portugal, and Malaysia crossed the threshold of 35 percent only in 2004. Poland, Spain, Greece, Qatar, Israel, and the Arab Emirates had 25 to 35 percent on line in 2004. Between 5 and 13 percent of China, Thailand, and Indonesia were on line in 2004, with India less than 5 percent (despite all the hoopla about outsourcing and "flat worlds") and the rest of south Asia less than 2 percent.

Another group that tracks Internet access is Nua Internet Surveys (nua.com). Its calculations are geographic, since distribution of the Internet is, lest we forget, by physical cables. Nua figures that 605 million people can access the Internet, and that 183 million of them live in the United States or Canada. But the rate of growth there has slowed, and equally large numbers of Internet users now live in both Europe (191 million) and the Asian Pacific region (187 million). More than 33 million residents of Latin America and more than 6 million Africans log on. In fact African Internet use has doubled every year since 1998. Starting from minuscule percentages, South Africa, the Cape Verde Islands, and Tunisia now have 7 percent, 3 percent, and 4 percent of their populations on line. Looked at another way, they are only two years behind where Great Britain was in 1997. There is an Internet acquisition curve, and many countries are just getting wired. They will have Internet use levels closer to those of North America within a decade, and the non-English content of the Internet will dramatically increase.

Meanwhile, according to Nielson/NetRatings, the Internet-using population of the United States itself held steady at about 165 million between 2000 and 2003. This is 59 percent of the U.S. population, the same percentage as Hong Kong, though not the highest in the world. Iceland leads the way, with almost 70 percent of its population on line, followed by Sweden (68%), Denmark (62%), and the Netherlands (61%). There are also almost as many Chinese Internet users (counting China, Hong Kong, and Taiwan) as there are Japanese users.

One thing to understand about these ratings is that they are estimates, and that different groups arrive at different numbers. Analyses tend to concentrate on growing and potentially lucrative markets. One such outfit, ClickZ.com, figures 135 million "active users" in the United States in 2005 and 7,000 Internet service providers (ISPs), the largest numbers in the world in both categories. By comparison, Canada had only 8.8 million "active users" and 760 ISPs, meaning that the average service provider has only 11,500 clients (the U.S. figure would be 19,000). By contrast, in 2005 Japan's 37 million active users connected to 73 ISPs (500,000 per outlet). Why should these ISP densities be so different? Well, cyberspace is not "flat." There are geographic paths, cyberspace difficulties, cultural legacies, and economic histories at play in each nation. Zook argues that e-commerce is grafted on to the stock of the economies before and after the dot.com bust. In some respects, the Internet follows paths as old as colonial transportation systems. ClickZ.com has other suggestive nuggets: Malaysia has nearly one-third of its population on line in some fashion, but Mexico a mere 13 percent. Both were colonies, both have oil money—why such a difference? And why does Bulgaria have a whopping 200 ISPs for only 1.8 million users? It is a center of fraud and cybercrime.

But even gauges such as Nua and Nielson/NetRatings are somewhat tainted by a Latin alphabet logocentrism. The newest domain names contain characters and marks (e.g., Thai and Russian) that web-crawling "spiders" are not good at measuring. There are now domain names using the 11,000 Chinese signs as well as other non-Latin characters. English-only browsers do not prioritize these sites in their searches, nor are U.S. computers and browsers usually set up to display these languages.

According to VeriSign, the domain registry, by 2003 English was already not the preferred language of the majority of Internet users, and it was the mother tongue of only 41 percent of them. Then sometime in the fall of 2003, English was eclipsed, falling below 50 percent of Internet content. That was also the point when the number of Internet users in western Europe and the number of Internet users in Asia surpassed those in the United States. At about the same time, VeriSign began accepting domain names in Latin, Greek, Cyrillic, Armenian, Hebrew, Arabic, Syriac, Thaana, Devanagari, Bengali, Gurmukhi, Oriya, Tamil, Telegu, Kannada, Malayalam, Sinhala, Thai, Lao, Tibetan, Myanmar, Georgian, Hangul, Ethiopic, Cherokee, Canadian-Aboriginal Syllabics, Ogham, Runic, Khmer, Mongolian, Han (Japanese, Chinese, and Korean ideographs), Hiragana, Katakana, Bopomofo, and Yi. So the Internet is rapidly becoming even more non-English.

As I mentioned in the discussion of English and "endangered languages," computing is helping to solidify these languages and to save others. Bosnian and Montenegrin are two languages being standardized and taught with the aid of computers. The African Languages Technology Initiative has developed a special keyboard for Yoruba, a tonal language, as well as voice recognition software. Microsoft now markets Windows, Office, and other products in Swahili. It has paid top scholars in Kenya to compile dictionaries and a glossary of 3,000 technical terms in Swahili. In Ethiopia research is underway to computerize Amharic, which has 345 letters; experts at Addis Ababa University have already come up with a text messaging system. Microsoft also plans to adapt its programs for Amharic, Zulu, Yoruba, Hausa, and Igbo. South African researchers are working on Afrikaans, Southern Sotho, Xhosa, Venda, and Tsonga. These languages would surely decline and eventually disappear without this buttressing by the Internet.

The fastest-growing segment of the Internet, however, is in Asia. By 2010 there will be 80 percent more users in the Asian-Pacific region than in the United States. China's 3721 and Foxmail, companies that we have never heard of, may become as large as Yahoo or MSNBC by 2010, and the percentage of English on the Internet will diminish even further. This won't be bad—English will still be instrumental—and both software and logocentrism will probably prevent English-speakers from realizing that they have been eclipsed for some years.

Do American Companies Dominate the World Economy?

In their detractors' imaginations, U.S. corporations are omnipotent forces that bulldoze meek foreigners into buying products, crush native competition, buy off governments, and extract money from underdeveloped countries with strong-arm practices. "By the intermediaries of the great, mostly American-based transnational or multinational corporations, a standard form of American material life, along with Northamerican values and cultural norms, is being systematically transmitted to other cultures," writes Jameson.[62] Critics take this supreme power as a given, designating it by the shorthand "TNC," for transnational corporation. The acronym allows ideological compression: all TNCs are assumed to be alike, regardless of ownership, origin, history, degree of internationality, field of business, sales, and number or location of employees. This blurs the realistic assessment of a company's everyday impact on life.

Among the many problems with this picture, the first is the measure of largeness. How do you measure the size of a company? Measured by market capitalization (the value of its stock), General Electric was the world's most valuable company in 1999, according to *Fortune*, worth $253 billion. Then CEO Jack Welch retired, and suddenly GE was worth only half as much. Measured by sales revenues, GE was only the tenth largest company that year. In number of employees, a measure of "size in *habitus*," it was way down the list (315,000 employees). There are different kinds of "large."

By most measures Wal-Mart was the world's largest company in 2003 and 2004. It led in sales revenue and was the world's largest employer, cutting weekly paychecks for more than 1.3 million people. This is multiple evidence that it is large, including size in *habitus*. But the second, third, and fifth largest employers in the world were Chinese: China National Petroleum (1,146,194 employees), SINOPEC (917,100), and the Agricultural Bank of China (491,000). Haven't heard of them? Critics like Jameson don't fret about their wages, working conditions, or power to standardize life. But any company that employs a half million people is not only "big" but also powerful. Some companies that Americans like to think of as "giant," such as Microsoft, aren't nearly as large in dimensions other than market capitalization or profits. Microsoft had about the same 2002 sales revenues as the cell phone producer Nokia. Is the Finnish company a "giant"? Neither company was among the one hundred largest in the world that year as measured by revenues, which is the ruler I'll use in this section.

The second problem is determining just what is "transnational." Definitions are of some help here:

> A multinational corporation (MNC) or multinational enterprise (MNE) or transnational corporation (TNC) is one that spans multiple nations; these corporations are often very large. Such companies have offices, factories or branch plants in different countries. They usually have a centralized head office where they coordinate global management. Very large multinationals have budgets that exceed those of many countries. They can be seen as a power in global politics. Multinationals often make use of subcontractors to produce certain goods for them. The first multinational, appearing in 1602, was the Dutch East India Company.[63]

General Motors, the fifth largest company in the world measured by 2003 revenues, meets these requirements, yet it isn't as "powerful" internationally as that rank would suggest. Most of General Motors' "power" lies the United States, where Americans buy its products. Its share of the European market was only 9.5 percent and falling in 2004, and those were

sales of its Saab, Vauxhall, and Opel subsidiaries. Even with these venerable assets, GM has lost money in Europe every year since 1999, and it has
outsourced so much domestic production to Mexico that it might not
deserve to be called an American company anymore (its Mexican subsidiaries sell more to the United States than GM sells in Mexico). The best-
selling car company in Mexico is Nissan, followed by VW. Are we permitted to have some reservations about GM's power as a TNC?

The "transnational" face of GM pales beside those of DaimlerChrysler
and Toyota, which have been the seventh and eighth largest companies in
the world for the past five years. Japanese auto makers in particular sell
far more outside of Japan than U.S. car makers sell outside of the United
States. In fact, Toyota passed DaimlerChrysler in the United States to
become number three in car sales in 2003, and if we subtracted light
trucks, it would be second, and Honda tied with Ford. The Japanese make
34 percent of the *autos* sold in the United States—and all three best-selling models. This figure would be higher except for quotas that Americans
call "voluntary export limits." The truth about many "large" U.S. corporations is that they are potent only in their home market, where quotas
and "antidumping" laws protect them from real global competition. They
become transnational to buy parts or to find cheap assembly. Selling
products in the United States is what they know how to do, but that
doesn't mean they can sell them abroad. We might think of them as "incidentally transnational."

The proper measure of a "transnational," in a discussion of corporate
power and globalization, ought to be its presence in foreign markets.
Volkswagen sells more autos in China than all U.S. car makers combined.
That's transnationality. The number two European car maker, Peugeot/
Citroën, sells more cars in South America than any U.S. auto maker. These
two firms are transnational auto companies.

If we look at the latest available United Nations Conference on Trade
and Development figures (2002), we find that its list of "The world's top
100 non-financial TNCs, ranked by foreign assets" contains twenty-five
Americans firms. This ranking uses a different measure, direct foreign
investment overseas. General Electric led in 2002, having edged past Vodafone of Britain. Ford was third, BP was fourth, and GM was fifth. Royal
Dutch Shell was sixth, Toyota seventh, Total-Fina-Elf of France eighth,
and Volkswagen was eleventh. But BP is the company that actually sells
the most abroad, followed by Exxon and Shell. Other interesting facts
emerge from this list: Vodafone is the TNC with the highest percentage
of its assets overseas, a whopping 80 percent. By contrast McDonald's has

only about 40 percent of its assets abroad, less than Germany's Bertelsmann, which owns a chunk of U.S. publishing. Toyota made $73 billion
in foreign sales, while GM made only $48 million.

Only about ten of the twenty largest U.S. companies (measured by
2004 revenues reported in *Fortune*) were significant TNCs when mea-
sured by their presence in foreign markets. That's down from twelve in
2002. Others, such as Fannie Mae, Kroger, Cardinal Health Systems, and
Berkshire Hathaway, are big at home but not players in foreign markets.
The companies legitimately called TNCs are Wal-Mart, General Electric,
Ford, GM, Citigroup, Procter & Gamble, Altria (formerly Philip Morris),
IBM, HP, Time Warner, Pfizer, and the oil companies ExxonMobil,
Chevron, and Conoco. These are also the U.S. firms listed in the UN
report's top fifty.

U.S.-based transnational companies "dominate" only a few industries, such as oil, financial services, aviation, and computing. Even there,
the world rankings would look very different without the U.S. home
market and the oomph it provides. For most U.S. TNCs, more than half
of revenues still come from their home market. As a result, most don't
export well—especially compared with the world's number two economy, Japan, or number three Germany. However, they are respected and
feared abroad because they are more driven by *profitability* than foreign
companies.

Among the fields in which U.S. firms do dominate, let's start with
entertainment, since according to Jameson, "whoever says the production of culture says the production of everyday life."[64] The following
figures are from *Fortune* magazine. The six largest entertainment companies in the world in 2004 ranked by sales revenues were:

Time Warner (USA)	$44 billion
Vivendi (France, now USA)	$29 billion
Walt Disney (USA)	$27 billion
Viacom (USA)	$26 billion
Bertelsmann (Germany)	$19 billion
News Corporation (Australia)	$17 billion

Missing from the list is Sony, the thirtieth largest company in the world
and usually the number one film distributor. It has 50 percent more
income than Time Warner and 250 percent more than Disney, but is considered by the *Fortune*'s list compilers to be a "diversified electronics"
company. If Sony's "entertainment" revenues were broken out, it would

rank with Bertelsmann and News Corporation. If hardware—from TVs to CD players—were included, then Sony would be the world's largest media company.

Even in this market where U.S. firms are so strong, foreign companies command more of the U.S. market than U.S. firms do of foreign markets. The world's biggest recorded music company is Polygram (Netherlands). The record division of Sony controls 19 percent of the U.S. market, and Vivendi Music (still French) accounts for 28 percent—these "foreign companies" controlled about half of U.S. music distribution in 2004. Such performers as Eminem, Limp Bizkit, Sheryl Crow, U2, and Shania Twain worked for the French, who accounted for 22 percent of world album (CD) sales in 2002.

A more pronounced area of U.S. dominance is aerospace and defense, where the only large rival is the European Aeronautic Space and Defense Company (EADS), maker of Airbus planes. Ranked by 2004 sales revenues, the top companies are:

Boeing (USA)	$63 billion
EADS (Europe)	$34 billion
Lockheed Martin (USA)	$32 billion
United Technologies (USA)	$31 billion
Northrop Grumman (USA)	$29 billion
Honeywell (USA)	$23 billion

The five American companies achieve their rank in this field by working for the U.S. government, hoping that the overpriced systems they sell will later prove exportable, with U.S. aid of course, to client states. Although it is a monopoly and receives EU subventions, EADS at least sustains itself in the civilian sector. Right behind this group is Bombardier of Canada (no. 8, $17 billion).

Among airlines, American logos are omnipresent, but they do not dominate the business. Few of them are profitable, but neither are their foreign rivals, most of whom receive government subsidies. Indeed, it is difficult to find an appropriate measure of size for an industry that has lost $32 billion since 2001. In late 2003 the market capitalization of low-cost carrier Southwest exceeded the combined stock market value of American, United, Continental, Delta and Northwest, making it by far the most valuable airline in the world. Of course, it doesn't fly outside the United States, so we can't call it a TNC. The decline of U.S. carriers in this group's rankings has been precipitous, while the French and British are actually making money. The top carriers, ranked by sales revenue are:

Lufthansa (Germany)	$18 billion (loss of $1 billion)
American (USA)	$17.4 billion (loss of $1.2 billion)
Japan Airlines (Japan)	$17.1 billion (loss of $.7 billion)
Air France Group	$14.5 billion (profit of $.1 billion)
United (USA)	$13.7 billion (loss of $2.8 billion)
Delta (USA)	$13.3 billion (loss of $.7 billion)
British Airways	$11.8 billion (profit of $.2 billion)

Note that the revenues of these foreign airlines ($61.4 billion) exceeded those of the top U.S. airlines ($44.4 billion) by 40 percent.

The United States also dominates the global securities business: the firms of Morgan Stanley ($35 billion in 2004 revenues), Merrill Lynch ($28 billion), Goldman Sachs ($24 billion), and Lehman Brothers ($17 billion) don't have much competition in stock trading. Stock ownership is not as widespread in other cultures, but where it is practiced, banks are often the agents. Some of them, such as BNP Paribas ($57 billion in 2004 revenues), are larger than any U.S. securities dealer.

Surely in banking the United States reigns supreme, no? Since *Fortune* now divides banks and financial services companies into categories, I use figures from 2002, the last unified list. The world's largest bank that year was Japan's Mizuho, followed closely by Citigroup. Japan and the United States each had three banks in the top fifteen. The Japanese had more assets than the Americans (due to inflated real-estate holdings), but the Americans made a lot more money. However, the European banks UBS (Swiss), Allianz (German), and Deutsche Bank (German) were all larger than the second largest U.S. bank, J. P. Morgan (which was seventh in the world). Banks we have never heard of, such as Paribas (France), HSBC (Britain), and ING (Netherlands) were larger than the good old Bank of America (before its recent acquisitions).

And when we look at some of those other financial services—such as life insurance—we find that the four largest life insurance companies in the world are Japanese; number one Nippon Life has revenues equal to those of Home Depot. Canada's Manulife is the second largest life insurance company in the United States, but no U.S. firm holds a comparable position in a country overseas.

Pharmaceuticals is another area in which the United States leads the world, but Europeans provide stiff competition, and the 2004 merger of Novartis and Aventus gave France the second largest drug company in the world. Ranked by 2004 revenues, the top drug companies are:

Pfizer (USA)	$46 billion
Novartis/Adventis (France)	$45 billion
Johnson & Johnson (USA)	$42 billion
GlaxoSmithKline (Britain)	$35 billion
Roche (Switzerland)	$23 billion
Merck (USA)	$22 billion
Bristol-Myers Squibb (Britain)	$21 billion

The business of computers and computer services is also an American strength, with IBM, Microsoft, Hewlett-Packard, and Dell in hardware, and EDS, Accenture, and Computer Sciences in the markets for business and specialty software. But more than half of IBM employees in 2004 were overseas, a figure that rose when it outsourced 4,730 highly paid jobs in late 2003 and then sold its ThinkPad business to China's Lenovo in 2004. Dell makes some computers in the United States, but more of them in Taiwan and Malaysia. Microsoft's Xbox game console is made by the Singapore firm Flextronics. Intel is the world's leading semiconductor maker, but with only 16 percent of the world market. It dominates the CPU market with its Pentiums, but Taiwan's Via Technologies, which produces the AMD and Athlon chips, has 40 percent of the CPU market. The rest of the list of firms producing chips for other purposes is Asian and includes Toshiba, Samsung, NEC, and Hitachi. Motorola recently abandoned the business, unable to compete.

In petroleum refining, BP (Britain) was the world's largest company in 2004. But following closely were ExxonMobil (USA) and Royal Dutch Shell (British-Dutch). The British and Dutch firms do more business by far in the United States than ExxonMobil does in Britain or the Netherlands. If there is a global superpower in oil, it is Great Britain. Total of France and ENI of Italy are also multinational refining superpowers, while Sinopec (China) and China National Petroleum dominate what will soon be the world's largest market. The SK corporation of South Korea and Repsol YPF of Spain both have revenues three times as large as Sunoco's. And when it comes to refining the crude, only five of the world's top twenty-five are American, a surprising fact given that the United States uses 20 to 25 percent of the world's oil. Foreigners sell us a lot of oil, and we sell them very little. The fourth largest refiner in the United

States, with 13,000 branded stations, is Citgo, which is owned by Venezuela and directed by its cantankerous president Hugo Chavez.

There are few other bright spots for America's "hegemonic" TNCs. Procter & Gamble is the world's largest "consumer products" producer. Coke is the world's largest beverage company, followed by Anheuser-Busch. But while Coke does an international business, Busch, which controls 50 percent of the U.S. beer market, is ridiculed by foreigner drinkers and whipped in the export market by Diageo. What is Diageo? you may ask. Diageo is a British company that is the world's third largest beverage company and purveyor of everything from Guinness beer to Jose Cuervo tequila. Other beers that we believe are North American—Labatt's, Rolling Rock, and Corona—are owned by Interbrew (Belgium). Miller Genuine Draft, Milwaukee's Best, and Jack Daniel's Hard Cola are owned by SABMiller of South Africa, which had 22 percent of the U.S. market in 2002. The rest of the big beverage companies are also foreign: fourth is Heineken, followed by Carlsberg, Brazil's AmBev, and Scottish & Newcastle. (Pepsico is considered a food company by list compilers.)

The final area in which the United States—the world's largest food producer—has a major presence is consumer food products. But even here Swiss and Anglo-Dutch firms rank one and two. Top food companies ranked by 2004 revenue are:

Nestlé (Switzerland)	$65 billion
Unilever (Britain/Netherlands)	$48 billion
Pepsico (USA)	$27 billion
ConAgra Foods (USA)	$22 billion
Sara Lee (USA)	$18 billion
Danone (France)	$16 billion

Nestlé, based in Switzerland, owns Ralston Purina and a flotilla of brands that most of us think are American. Nestlé's profits even grew 13 percent in the dismal business year of 2001. When it comes to coercive behavior, Nestlé leads the way, providing milk formula free to nursing African women. Its rivals are Groupe Danone of France and Unilever of Britain, while Pepsi and Sara Lee remain less known outside the United States. The major distributor of food products in the world is Carrefour of France ($80 billion in 2004 revenues), which operates more supermarkets than any U.S. company and is the number five grocer in China.

What about Wal-Mart, the world's largest retailer? It is a huge and powerful company. Some of its overseas operations (Mexico) are successes, perhaps aided by the familiarity of migrant workers with U.S. stores. But after six years in Germany it was still losing money in 2003. Its modest effort in Japan was completely outflanked by the Aeon Corp, Japan's second largest supermarket chain, which opened thirty superstores before Wal-Mart could open one. It was rumored in 2005 to want to buy Daiei, a bankrupt Japanese retailer, to gain some foothold in the second largest economy. But a lack of local cultural and marketing knowledge clearly hobbles the world's largest company as it attempts to expand. Meanwhile Chinese consumers flocked to enormous urban malls that offered more shopping variety than Wal-Mart's big box model did.

Those are the fields of "dominance" of American TNCs. The largest employment-services firm in the world is Adecco of Geneva—also the largest outsourcing company. The largest wireless phone companies are in Asia (China Mobile, China Unicom, NTT DoKoMo) and Europe (Vodafone, T-Mobile). Two of the four largest PR firms are European (WWP Group of Britain, Publicis Groupe SA of France). Michelin (France) and Firestone/ Bridgestone (Japan) are larger than Goodyear internationally, and they own nearly as much of the U.S. market as the leading U.S. tire maker. One-third of the shoes that Americans wear are made in the Guangdong province of China. Mexico and China produce more cement than any U.S. company. The world's leading chemical companies are BASF and Bayer, both Swiss (followed by Dow and DuPont).

In the worlds of grocery retailing, banking, and fashion—all potent areas of cultural transmission—if someone overseas looks up at a store logo, it will *not* be A&P or Albertson's. The world's top grocers in 2004 revenues are:

Carrefour (France)	$80 billion
Royal Ahold (Netherlands)	$63 billion
Metro (Germany)	$61 billion
Kroger (USA)	$56 billion
Tesco (Britain)	$60 billion

There are no Kroger stores outside of North America, but Royal Ahold owns the Tops and Stop-n-Shop grocery chains in the United States. Ahold produces its own house brands from corn flakes to catsup, exporting them to the United States. After acquiring fifteen more small chains in 2001, its sales rose 46 percent in 2002—it was by 2004 the largest grocer in the eastern United States. There are Carrefour groceries all over Europe, in Taiwan and China, and even in Argentina.

Among producers of electronics, only one U.S. firm (Tyco, no. 8) ranks in the top ten, and only three among the top twenty. Siemens of Germany is the world leader, but the next seven, and nine of the top twenty, are Japanese. We know their names. Can a nation be a world leader and *not* make electronic products? There are no televisions or CD players made in the United States. Can a nation be a "transnational media power" and not make delivery systems?

The world's top engineering and construction company is not Vice President Dick Cheney's alma mater Halliburton but the French firm Bouygues, which is also a force in European telecommunications. Number two is the French firm Vinci, followed by *nine* Japanese firms. Obviously the French and Japanese do a great deal more in the way of designing shopping centers, airports, and apartment buildings—influencing daily life—in the rest of the world than the United States does.

What about the rapidly growing and life-structuring business of telecommunications? Who dominates the world of cell phones and land lines? The world leader is Japan's NTT, whose revenues surpass those of second-place Verizon. The most promising business of Verizon—cell phones—is half-owned by Britain's Vodafone PLC, which is the world's fourth largest telecom. Vodafone dominates not only the British Isles but Germany and is the world's premiere cell phone provider, with a controlling interest in Japan Telecom. Deutsche Telekom is the world's third largest, AT&T was fourth (in 2004), and France Telecom is fifth. But neither of the U.S. telecoms is international. And they have guaranteed that they may never be by adopting a transmission standard used only in the United States. On the equipment side, Nokia of Finland battles it out with Samsung.

In the crucial field of scientific and control equipment, Fuji of Japan is the world leader (2004 revenues of $23 billion). Can a nation with a large landmass but no world-class railroad companies be considered a world power? The world's largest railroad companies are German, French and Japanese, Japanese, Japanese. Union Pacific just makes the list. Dominance in railroads assures that a nation will produce rail cars and systems, which export extremely well, because developed nations need railroads and subways. Dubai is having a railroad built: Mitsubishi and Kajima are building it.

The world's largest publisher is Lagardère Groupe of France, followed by Dai Nippon Printing and Toppan Printing of Japan. The world's largest metal maker is Mittal Arcelor, which has plants throughout Europe, Asia, and South America. Nippon Steel is second, Norsk Hydro of Norway third, and JFE of Japan fourth. Then comes Alcoa, looking over its shoulder at Baosteel of Shanghai. In specialty metals, America's Alcan sells only half as much as South Korea's POSCO consortium.

If an underdeveloped nation wants to build a sugar factory or an airport, whom can it call? Thysen Krupp of Germany (2004 revenues of $39 billion). At construction sites, the names on the equipment are mostly Asian, such as Komatsu. Japan's Mitsubishi Heavy Industries and Kawasaki make a good deal of it. The United States may be a great farming nation, but France's Alstom was a bigger farm equipment company than John Deere in 2004. Down on the farm there are a lot of European and Japanese machines these days.

Well, what about fast food? In 2004 McDonald's worldwide revenues were second to those of the number one Compass Group of Great Britain. The third largest firm is French, Sodexho Alliance. Together the British Compass Group and the French Sodexho Alliance have revenues 20 percent larger than the two largest U.S. fast-food purveyors, McDonald's and YUM brands (Taco Bell, KFC, Pizza Hut, etc.). So who is selling the world on fast food? Still, the American logos trump those of their British and French peers. In fact, even the American-produced film *Super Size Me* (2004) mistakenly identified both Burger King and Sodexho as American companies.

There are also giant corporations of a genus unknown in the United States, called "trading companies." Most people would agree that the biggest banking and financial services company in the United States—Citicorp—is *big*. But in 2002 the Mitsubishi Trading Company (revenues of $109 billion) was far bigger. Mitsubishi and Mitsui, the second ranking trading company, had combined revenues greater than those of General Motors. Put the trading companies Mitsubishi, Mitsui, and Itochu (no. 3) together, and we have sales greater than those of number one Wal-Mart. There simply are no U.S. companies in this category, but they are *sui generis* global, and they reside in Singapore, Seoul, Hong Kong, and Tokyo. But categorizing businesses is a cultural choice.

One final category in which foreign firms are, oddly enough, larger than American firms is energy companies. Many foreign ones are huge, state-owned monopolies—not very profitable, but very large. The "State Grid" electricity company of China is the forty-sixth largest company in the world, Electricite De France the sixty-second, E.ON of Germany (water and natural gas) the sixty-seventh. Suez, Gazprom, and Veolia (formerly Vivendi) are other unfamiliar names, but they are all larger energy sellers than Duke Energy, America's biggest power company (204th in the world).

THE *HABITUS* OF THE TNC

It is useful to remind ourselves of the myriad ways in which even true TNCs *must* be local. Surveys show that only 2 to 10 percent of TNC employees abroad are expatriates from the home office, mostly the top executives. Of Carrefour's 28,000 employees in China, only 79 were French in 2005. Pepsi and Apple sent managers to the Czech Republic in the 1990s, but when I visited in 1993 they were already bringing them home, their positions filled with newly skilled Czechs. One Prague headhunter said that "Foreign firms are already trying to get rid of the expats."[65] Local executives were less expensive, and they understood their culture and its needs better.

In 1969 there were just over 7,200 transnational companies in the world, with 60 percent of them headquartered in the United States according to *Global Inc.* By the year 2000 there were 63,000 such companies, a ninefold increase, but fewer than 30 percent were headquartered in the United States. The rest of the world had become transnational too. Some of these companies are bigger than others, of course. Of the 500 largest TNCs, 185 are headquartered in the United States, 126 are based in the European Union, and 108 are located in Japan.

But even the most insulated foreign executives *must* eat some local foods, shop in local stores, walk the sidewalks, drive the streets, use the dry cleaners. They may work ten hours a day, but the other fourteen are spent in local contexts, which affect everything from the language they speak to the clothes they wear (do they dress for Kuala Lumpur or for St. Petersburg?) to the way they sleep (futons or duvets?). Their children may attend international schools that teach in English, but they will also learn the local language and mores from the children of the local oligarchy. If they have to be abroad, transnational families are known to prefer to stay in one country, rather than moving. They put down roots in São Paulo or Singapore. The *Wall Street Journal* noted with a tone of alarm in 2004 that this type of transnational family (think of Carlos Ghosn of Nissan, or Roberto C. Goizueta of Coke) was becoming common at the top of U.S. TNCs, while the pure products of America were rarely tapped to lead foreign corporations. Instead of going abroad, "Americans think about developing their careers in America, where the playing field is very large," says Roger Brunswick, a New York consultant.[66]

Most foreign TNC employees, when they leave one TNC, go to work at another one locally. These are not top executives, but people who also become rooted in a place. In Tokyo and Osaka there are thousands of expatriate Americans, Australians, Filipinos, and Indians who work for whatever TNC offers the best pay and conditions. This is true in Hong Kong, Helsinki, and Frankfurt, too. They move from Procter & Gamble to Carrefour to Royal Ahold NV—these companies all sell groceries and need similar skills. But whether American or Japanese or French or Filipino, these workers *live* a local *habitus*. No matter where the home office is, the TNC's building, cleaning services, holidays, caterers, start time, offices and doors and desk sizes, and secretaries are all local. Sometimes the only uniformity from branch to headquarters is the logo over the door. Procter & Gamble is headquartered in Cincinnati, Ohio, a somewhat provincial midwestern city. Its offices and employees in Paris and Kobe are very different. The *habitus* infiltrates the local P&G offices. Of course, TNCs work hard to keep their foreign offices in synch with the headquarters. But when Procter & Gamble's Asian division, based in Kobe, decided in 2000 to alter its stock numbers (and hence its bar-coding), it took six months to get all the branches, which stretched from Manila to Bombay, to change their software to the new standard. Kobe employees told me that the Cincinnati home office was powerless to speed up the pace.

Some critics tell us that the locale of the parent company is all that matters: its ideology somehow passes abroad, while profits return home. But others argue that the locale of the research center is most important, because it is the intellectual capital of the company. Still others say that physical plant and production are the locus of power. What would they say about Sony's Entertainment Division? The home office is in Tokyo, the "research" in Los Angeles, New York, and London, and the production of CDs and tapes in Taiwan and China. What about Chrysler? It is headquartered in Germany, but its products are designed in Michigan, with many parts produced in Mexico. Rupert Murdoch began the News Corporation in Australia, has his most valuable assets in the United States, and is headquartered in Britain. Transnational corporations are often so decentralized that they resemble logistical confederations.

Nor are transnational companies capable of pulling up stakes quickly. Honda may detest Ohio's worker's compensation laws, but it isn't going to dump its $123 million investment in a new plant at Marysville, Ohio. Procter & Gamble distributes consumer goods to stores worldwide, so it might seem able to shift its operations around, but its supply chain is too complex and too carefully calibrated. As Robert J. Antonio and Alessandro Bonnano have stated, "TNCs are not 'placeless' or 'deterritorialized' phenomena."[67] Many American TNCs, in fact, are charmingly provincial, anchoring their headquarters in the places where they began. Wal-Mart remains in Benton, Arkansas, and Microsoft in Redmond, Washington. McDonald's is still in suburban Chicago, and Coke in Atlanta. Not many TNCs have pulled up stakes, moved elsewhere, and prospered. This has something to do with the way a specific focus is achieved in a *habitus* and a widespread suspicion within companies that their formula might not work from elsewhere.

Another charge against TNCs is that they dictate to foreign governments. This notion is popular among academics who think that corporations are replacing nation-states. Were they watching when Coca-Cola tried to buy the French beverage maker Orangina in 2000? The French government just said no. Orangina is sold in every bar and restaurant in the country, and the government reasoned that Coke would have an instant, nationwide delivery system for all of its products. End of story. In 2001 the European Economic Community said no to General Electric, which wanted to buy Honeywell, *another American company*, because Honeywell had too much business in the European Union. In 2004 Japan booted Citibank's private banking group from the country because it didn't like its business practices. It also threw out Crédit Suisse in 1999. Mexico and Canada have vetoed U.S. companies' moves in numerous industries, from soft drinks and trucking to magazines and pharmaceuticals. Canada's refusal to allow U.S. magazine editions to be sold there is notorious. Other national governments are even stronger. The head of one of Russia's largest companies, Mikhail Khordorovsky of Yukos Oil, was sentenced to nine years in jail when he resisted President Vladimir Putin's attempt to dismantle his company. Nation-states are not disappearing, as TNCs would be first to explain to academic theorists.

Make no mistake: some American companies, such as Exxon and Wal-Mart, have tremendous international clout. And many are highly profitable, which their rivals fear as much as their presence. But they have lots of competition from local companies, which usually know the local markets better, and from other multinationals, which have comparable advantages of scale and cheap sourcing. Then there are nation-states, which have more clout, as the GE and Coke cases demonstrate, than any TNCs. U.S. multinationals may play hardball, but they don't have the game sewn up.

Seeing Ourselves Everywhere

Globalization looks disproportionately "American" to Americans because when they travel they readily recognize familiar products and components of their own lives. They recognize their *habitus*. Critics in particular seem prone to make the leap from recognizing the familiar to
declaring it the pattern. Why is this so? Why does so much analysis of
globalization insist, for example, on assigning a transcendent level of
meaning to logos? Why does the world beyond American shores appear
so deplorably "Americanized"?

It is not enough to say that these are mistaken or partial perceptions.
These are misperceptions that arise from a particular set of cultural circumstances, which we need to get beyond in order to see globalization
for what it is and what it is not. The American traveling abroad tends to
see pernicious globalization everywhere, but at home the same person
regards the introduction of sushi or French bread locally as a pleasant
increase in culinary choices.

There is a great deal of resistance overseas to globalization, some of it
ideological, but much of it deeply embedded in local cultures. If we sum-
mon to mind the scene that opened this chapter, the logos of Nishinomiya, we will recognize that what we "see" is that part of modernity
that we already know, but defamiliarized by its new, foreign context. What
happens is that we grasp first the signs that we recognize in our effort to
make meaning. Thus Burger King = familiar food. Then inside the Burger
King, there are Japanese people! Asian-flavored sauces! But we assimilate
them to what we recognize. Now let's reverse the situation. Let's pretend
we are Japanese entering a sushi shop in the United States. "Sooo many
Americans!" is the first reaction of my Japanese friends. "But what is this
Philadelphia roll?" And we answer back, "Of course, Americans like sushi
too—we've been eating it for some time. The Philly roll uses cream
cheese, and it's not bad!" Our answer reveals that the "new" is everywhere
naturalized and adapted, chosen and received, by consumers, by "us." The
cultural matrix that dictates how this happens is the subject of chapter 2.

2 The Resistance of the Local

On Sunday morning most of the French who live in old Avignon pay a visit not to church but to Les Halles. In this cramped market at the center of the city, they meet their neighbors, co-workers, their dentists and doctors, friends in from the country, their children's teachers, and, later on, the priest. The food sold here is mostly local—mushrooms and cheese from the Luberon, truffles and some boar in season, seafood from Marseilles, mounds of fruits and vegetables from the Rhone valley. The few imports are fruit from Italy, Israel, and Morocco, English jellies and canned goods. There are no organically modified foods, no Coke, no fast food, no logos, no chains, no advertising, and few tourists. The language of the vendors is French, spoken with *tu-vous* distinctions and the sing-song intonation of commercial custom. Many customers reply in Provençal, a few in Occitan. I would not be understood in English. Many stall owners are third or fourth generation, and at least one traces his lineage back to a *fournisseur* of Clement, one of the "black" popes.

There are comparable scenes all over the world. Once we can *see* this local resilience, resistance to globalization is everywhere. Yet critics like Jameson see not resilience but fragility: "Each national culture and daily life is a seamless web of habits and habitual practices, which form a totality or system. It is very easy to break up such traditional cultural systems, which extend to the way people live in their bodies and use language, as well as the way they treat each other and nature. Once destroyed those fabrics can never be recreated."[1] In fact, it is not "very easy to break up such traditional cultural systems." And his colleague Sherif Hetata's notion that "the spread of global culture is the corollary of a global economy" is just a bad metaphor.[2] Culture does not work by corollaries—it is not math—and it is only partly rational or economic. Culture is local. It is incredibly durable, and I detail some, but hardly all, of the reasons why in this section. One of the major reasons for local resistance, as Dusan Kecmanovic has explained, is that the global economy itself raises insecurities that intensify regionalism, ethnocentrism, and nationalism—in short, the global economy intensifies local culture. Even if nation-states disappeared, Kecmanovic argues, ethnoregionalism would endure. This is not hard to understand: from infancy onward, customs of food, language, gender, use of space, education, work patterns, cleanliness, thrift, religion, racism, honesty, and regard for authority overwhelm the individual. These forces truly create us. Any individual who could escape the formative grasp of his or her early acculturation to adopt "globalized culture" would be quite extraordinary.

Most of what we call "culture" is formed in infancy. Mountains of scholarship exist that document the predisposition for early cultural imprinting; these show that infants acquire much of a culture before they can speak. As children, they learn foods, languages, spatial systems, gender relations, and family structure long before they are ever exposed to anything commercial, much less "globalized." They are initiated into educational patterns, religion, a culture's approach toward work, toward honesty, and toward authority. Attitudes toward race, foreigners, mechanization, and migrants follow. As sociologists Geert Hofstede and Ronald Inglehart have shown in large pan-national studies, these cultural attitudes have proved extremely obdurate to globalization. Let's examine just a few of these features.

Language

Language acquisition is the subject of an immense literature, but it can be bowdlerized by stating that there is no language like *la langue maternelle*. The language learned in infancy is not only the one learned best, but also the one that structures an individual's expression most importantly, even in speaking other languages. Only professional translators and those who have grown up truly bilingual have full command of a second language. To give but one example, the subtle shift in English meaning caused by a change from the indefinite article *a* to the definite article *the* escapes the first notice of nonnative speakers even after decades in the United States. Comparable difficulties exist in the other languages that I know (French, Spanish, German, and Japanese). Most multilingual people speak their second language in a patois, missing prepositions or liaisons or postpositional markers. The rising and falling intonations of Asian languages, the reshaping of the interior volume of the mouth required in Arabic and some African languages, the various *th* sounds of English—these escape even university professors who spend decades acquiring their second or third languages. To get the accent right, researchers say, we must learn a language before twelve. My point is simply the primary-ness of *la langue maternelle*. Second languages are spoken with an accent, a limited vocabulary, and deficiencies in the cultural weight of specific words and concepts.

La langue maternelle structures experience in a particular way through its vocabulary, word order, verb tense system, and method of pluralization. I am not dusting off Sapir and Whorf, Chomsky, or any other crusty linguist here. Languages have these structures to habituate speakers to the world views of the concomitant cultures. Japanese, for example, has no formal future tense, yet the Japanese obviously think about the future—they use extra present tense and future time markers. But they do not have anything comparable with future anterior or future conditional tenses in English, modes that make hypothesizing about future possibilities much clearer and easier. On the other hand, a pronoun-addicted language like English lacks the necessity of close attention to group context, which Japanese requires. If everyone seems to be on the same page in Japan, it's because everyone has to pay close attention to understand anything. The languages vary according to the cultures they express. The Romance languages gender every noun, and while one wouldn't want to make too much of that these days, that has an effect on the speaker's world view. English lacks the sense of agentless action connoted by reflexive verbs in Spanish. Japanese has levels of politeness that are almost parallel languages, dwarfing the *tu/usted/vosotros* distinctions of Spanish, to say nothing of a certain language in which everyone is "you." Recent attempts to eliminate this *keigo* Japanese, on the theory that American-style "you"-speak would lead to more innovation, have proved mostly futile. Workers interviewed by the *New York Times* spoke of sweating profusely when forced to address the boss as "Kubota-san" instead of "President Kubota-san." Japanese distinguishes between subtly different states by a plethora of nouns—rice in the field is *kome,* rice in the pot is *gohan,* and foreign rice is *raisu.* While the Japanese language doesn't determine how the Japanese people *think*—at an abstract level they know that rice is rice—the language provides a particularly close embrace of small differences in states of being, which is part of the culture.

So the idea that English or any lingua franca is going to displace *la langue maternelle* and its value-laden embrace of reality is just nearsightedness. If all parents in the world spoke English to their children from day one, communication between different nations would still take place only in a *pidgin* English, because other cultural factors are so deeply implicated in human communication.

Ways of reading a language, for example, are also embedded early. Japanese is traditionally read vertically, right to left. Some Chinese and all Hebrew are printed and read from right to left. Some languages are read from the back of the book to the front. There was an uproar in 2002 when Chinese newspapers in the United States changed from the traditional system to the left-to-right, horizontal system used in mainland China. Counting and measuring systems have remained impervious to globalization. Why do Americans, the most globalized people on earth, refuse to give up measurement by Fahrenheit, gallons, miles, and inches? The French retain their nonsequential house numbering and *bis* addresses. These systems have their own histories and pace of development. In Austria stairways, as well as buildings, are numbered. The Japanese *ku* and *cho* system (streets are usually not named) defies Western logic. Then there are the twenty-plus shape and type counters that Japanese suffixes to countable objects. Money systems are similarly integrated into local languages in myriad slang, and woe to the next bureaucrat who introduces a two-dollar bill or a Susan B. Anthony dollar. Some Asian and Middle Eastern nations use different calendars.

Global technologies have also put the means of promoting and preserving *les langues maternelles* within reach of every culture. As noted in chapter 1, radio and television in local languages have become the norm. Audiotapes, videotapes, and CDs offer children's fare in languages from Inuit to Farsi. These are easily copied and passed from household to household. *Sesame Street* broadcasts in a gamut of tongues from Catalan to Swahili and Cantonese. Photocopying and cheap printing permit children to read local lore and textbooks in languages that did not even have written forms fifty years ago. The Soviet Union abolished the Chuvash language fifty years ago, but due to technology and ethnocentrism it is today spoken throughout the republic of Chuvashia.

Communicative Distance

On the island of Bali, an infant's feet never touch the ground until the child is several months old. Babies are carried constantly, by a variety of community members. In Senegal too, infants are passed around and tended by an extended family that may number eighty members. In the United States, there is no carrying around at all. Children as young as twelve months are dropped at day-care centers at 7 a.m. and retrieved at 6 p.m. In France, where single mothers abound, grandmothers are still very active in child-care. In Japan, an estimated 1 million young men, dealing with emotional or academic defeat, have shut themselves into their bedrooms and refuse all communication, a phenomenon called *hikikomori*. They sometimes stay there for years. Americans feel their "personal space" invaded when Spaniards and Italians move up close to talk.

These are all examples of "communicative distance," which is the intimacy or formality of speech context, the friendliness or aggressiveness of speech acts, and the expectation of speech frequency. Among other things it governs the expectation that one speaks differently to different types of people or the same way to all people, a cultural category called *particularism* by sociologists. Communicative distance is established at the same time as speech, and it endures throughout a lifetime. Even after mastering a second language, most bilingual adults are unable to achieve its communicative distance without conscious modeling and practice. One would have to study, as did Davide Sesia, the director of Prada in Japan. He learned that when he negotiates, maintaining a humble posture, proper distances, and delaying Japanese verbs to the ends of sentences "gives me extra time to react to indirect facial expressions and even to reverse the entire meaning of a sentence."[3] By studying Japanese communicative distance, he has managed to secure great store locations without the down payments usually required of foreigners.

Examples of communicative distance are legion. While many cultures shake hands, nowhere outside the United States does its back-slapping, chest-bumping, high-fiving macho male behavior appear. In fact, men do not embrace nearly as much in cultures outside the United States, although they may kiss intra-*familia,* as in France, or hold hands, as they do in much of Africa. Kissing is one of the most overt French communicative distance habits; another is the use of *tu* versus *vous* address. Japanese communicative distance is much different. Physical touching exists only in the immediate family, between lovers, and in contact sports. But Japanese infants are touched and handled a great deal, mostly by their mothers. Rather than learning to wash themselves, children are washed until school age. They learn about human separateness through the ideology of cleanliness, as mothers explain that contact spreads disease. So while Japanese are acculturated to close proximity with others, they avoid contact except with intimates. On crowded subway cars the Japanese try not to touch anyone else; this is one reason those famous "packers" are needed. Even drunk Japanese salarymen tend to maintain appropriate communicative distance. On the other hand Poles and Russians, who are equally reserved in daily public life, discard traditional communicative distance when drunk. In France it is considered proper form to pause before leaving a bus to say "Au revoir" and "Merci" to the driver, but unassimilated North African immigrants don't, which causes social friction.

There are other cultures where immediate intimacy is the norm. Despite population densities similar to Japan's, strangers in Thailand are touched, taken into close speaking proximity, and addressed familiarly. In West Africa boys walk with their arms around each other. Visitors to Italy notice the scant distance between speakers immediately. In Japan, they notice that the distance has been lengthened. And the bowing, which is an art unto itself! In the United States men gaze at women, but in Brazil the women gaze at the men. These features in aggregate have thousands of small repercussions for daily life, and they are endlessly transmitted by radio, television, and film in a way that confirms and reproduces them for children.

A failure to understand communicative distance has dogged the U.S. efforts in Iraq. " 'Welcome' is probably the most widely used English word here," wrote John Tierney in the *New York Times*. "Even if his kitchen has just been destroyed by a car bomb, an Iraqi host will apologize to a visitor for not offering the ritual cup of tea. But Americans often do not know how to reciprocate politely. They routinely offend Iraqis by plunging into business instead of paying respects to the host and asking questions about his well-being and that of his family."[4] Their translators tell the Americans that "You can't ask that question here." To spare everyone embarrassment they simply ask a different question in Arabic.

Communicative distance shows no sign of changing because of globalization. In fact, a great deal of newly created wealth is dedicated to maintaining it. Ronald Inglehart cites Saudi businessmen who spend small fortunes building relationships with *individuals* from Western companies. They will deal only with those people, as their traditions dictate. Their American counterparts experience extremes of isolation and physical contact, moving to suburbia to be alone but driving to urban sports stadiums and discos, where they are in intimate contact with hundreds or hundreds of thousands of people. Riders commuting to and from New York City on the Long Island and New Jersey railroads will stand or sit in the aisles rather than use the middle place in three-across seating. The Japanese build capsule hotel rooms, where for thirty to fifty dollars a night they can cocoon in a one-piece plastic berth measuring 150 by 200 centimeters—and remain closer to work. Americans prefer to commute long distances to sleep in their own enormous beds. The British use their new wealth to travel abroad extensively, but the French do not—60 percent of Frenchmen have never left France. In the United States and France, children usually leave home by their early twenties, but in Italy more than 33 percent of those thirty to thirty-four years old live with their parents, preferring home comforts and their mothers' cooking. Wal-Mart in Japan cannot get employees to adopt its American guideline of asking any customer within ten feet "How can I help you?" That is so far beyond normal communicative distance in Japan that it's like shouting across a football field in the United States. Besides, in Japan clerks should be passive, not speaking until spoken to.

Cultures also deploy new technology to reinforce traditional communicative distance. Cell phones are an example. Japan and Finland have more subscribers, the highest usage rates, and the most sophisticated cell phones in the world. But these phones replicate and intensify some traditional aspects of Finnish and Japanese communicative distance. Movement in trains, buses, and cars is a constant of modern Japanese life, but phones permit the Japanese to be in contact with family, business associates, friends, and lovers while in movement. They replicate the information sharing, consensus building, and obligation creation that were achieved through social visiting, gift exchange, *omiyage,* and consultation in the past. A young woman thinking of buying a computer, for example, calls her father and her friends for recommendations and ideas, rather than checking the ratings, as an American might. Anecdotal information and recommendations return to the young woman, reduced to a consensus, as well as the names of people to contact. But the highly educated Finns, almost laconic on the phone, rarely use them for consensus building and make far fewer calls per capita than the Japanese or than neighboring Swedes and Norwegians.

Communicative distance has been measured by social scientists on the scales of universalism versus particularism (Inglehart), as low context versus high context (Edward Hall), and earlier as U-type and G-type (Kurt Lewin). These measures distinguish between cultures in which people engage with others only in specific areas of life or only at single levels of personality, as opposed to cultures in which people communicate across all areas of life and personality (see fig. 2). Americans are very specific, very low context. In the workplace they may be on a first-name basis with the cleaning crew, but they don't socialize with them. The

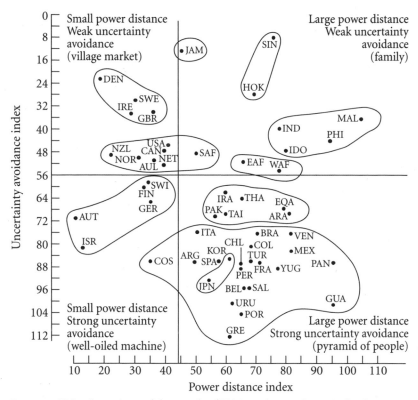

FIGURE 2. Using interviews of thousands of IBM employees, Geert Hofstede mapped the "uncertainty avoidance" and "power distance" of people in fifty nations and three world regions. He showed not only how enduring basic cultural attitudes can be (the United States, Canada, New Zealand, Australia, Britain, and Netherlands cluster together in the upper left quadrant), but also how different even prosperous countries of the same region can be: Japan appears in the lower right quadrant, while Singapore and Hong Kong are in the upper right.

SOURCE: Geert Hofstede, "Uncertainty Avoidance and Power Distance in 50 Countries," in *Cultures and Organizations: Software of the Mind* (New York: McGraw-Hill, 1997), 141 (fig. 6.1).

French and the Germans are medium specificity, high-context cultures: they greet the *femme de ménage* enthusiastically on the street or in a bar, but remain *vousvoyer* in address and restricted in topic range. Japan is a high-context communication society, where almost everything is already implicit in the social context. One hardly ever talks to the cleaning people. As Hofstede notes, Japanese business contracts can be half the length of American contracts, because both sides already understand the "what-ifs." Until 2000 there were only a few thousand lawyers in the whole country, though more are now being trained, to deal with intellectual property and piracy.

Food

"It is probably in tastes in *food* that one would find the strongest and most indelible mark of infant learning," Bourdieu writes, "the lessons which longest withstand the distancing or collapse of the native world and most durably maintain nostalgia for it. The native world is, above all, the maternal world, the world of primordial tastes and basic foods."[5]

These tastes begin at the breast. In Norway 99 percent of mothers breast-feed their children initially. In the United States the figure is 70 percent, in France only 50 percent. In some parts of Africa and Asia, the figure is less than 30 percent. French infants are fed yogurt and other manufactured milk products at an early age, but in Japan, as Ruth Benedict long ago reported, infants are nursed to eight months, sometimes until "they can understand what is said to them," and then moved right to the family table where they "are fed bits of food."[6]

But that's the end of milk in Japan, as children are weaned to rice water and then a rice gruel called *okayu.* Japan has many forms of rice: rice cakes, rice candy, rice puffs. The young are encouraged to focus on the texture and mouth feel of rice. "Good" rice has granularity but also sticks together for ease in handling. It is never mushy, never chewy. Good rice has a mild nutty flavor, a slight aroma, but no aftertaste: it invites you to keep eating it. John Dower writes in *Embracing Defeat* that rice had to disappear completely during World War II before the Japanese would consider eating barley or potatoes. Even expatriated Japanese in the United States start their babies on rice gruel at two months and whole rice at five months. In contrast, anthropologist Sidney Mintz tells us that Americans have traditionally gotten two-thirds of their carbohydrates from potatoes and wheat.

French children also get their culinary indoctrination early. *Blé,* a finely milled form of wheat, is sometimes added even to infants' bottles, to provide the ballast to get them to sleep through the night. Grocery shelves are crowded with *blé* products produced by Nestlé, Bledina, and HIPP, such as Bledi lait, P'tit Biscuit, and Lait et Céréales Cacao. "Gôuter aux flocons de blé" advertises one product, while another claims to mix *blé* with "p'tit mouliné légumes a la Provençal." Yogurt is introduced at five to six months, and children learn to respond to a wide variety of milk products, from flavored yogurts and *crème fraîche* to those cheese triangles of La Vache Qui Rit. Bits of French bread are torn off and fed to infants as soon as teeth appear; in fact, bread crusts are a favored teething tool. In contrast to the situation in the United States, there are relatively few protein-rich French baby foods: no eggs or meat or fish, and only a few with chicken. The French infant grows up with *blé* and moves to the *baguette*. Sour tastes, ranging from olives, pickles, and sour chewing gum to wine (served in diluted form at the family table) arrive at around age ten. Children are introduced to the seasonal rotation of foods at home and in the school *cantine*, where there is usually professional cooking, as well as dressed salads and *sauce blanche* (although lots got left on the plate in my children's experience).

Infants off the bottle in the United States, on the other hand, graduate to fruit juices and soft drinks. Sugar, in short, which as Sidney Mintz has argued, is a kind of populist methamphetamine. Early foods include applesauce, strained plums or apricots, sweetened carrots or squash. Early protein, from chicken to beef, is also sweetened. The "mouth feel" of U.S. infant foods is made smooth not only by straining and blending but by emulsifiers. These two sensations—smoothness and sweetness—become the base of American eating habits, which recent studies show are set before age two. Two-year-olds already eat 30 percent more calories than they need, and 30 percent eat no vegetables. One-quarter of U.S. children over age one eat hot dogs, sausage, or bacon daily. French fries are the most common vegetable consumed by American children fifteen months and older. Lives there an American three-year-old who has not tasted ice cream, birthday cake, or a hot dog? Study on study has shown that Americans don't like low-fat products. Innovative new foods like water buffalo, with its low-fat, low-cholesterol meat and rich milk, consistently fail in the U.S. market.

A child growing into a culinary system learns to use utensils. We are told that one-third of the world eats with knives and forks, one-third with chopsticks, and one-third with bare hands. This is no small factor in local food culture or in the persistence of local taste. "Subsystems usually set the terms against which these [food] meanings in culture are silhouetted," writes Mintz.[7] Certain foods are more easily eaten in these modes, and we become adept at a particular technique. Most Westerners refuse to try or are grossly incompetent with chopsticks: the Japanese and Chinese don't like each other's chopsticks (wood vs. plastic). Rice that sticks together is favored in chopstick and hand-eating cultures. Peas and applesauce are not. Indonesians make do with a fork and serrated spoon.

By eight or nine a Japanese child has eaten a variety of unsweetened, unemulsified rice products such as rice cakes, *mochi,* and *sembei,* with a strongly contrasting spice system: the bitter, sour, astringent, and hot tastes of *umeboshi, nori, tsukemono,* and *shichimi.* There is a special vocabulary for describing these flavors, such as *shio-karai* (salty-hot). Asian children are also introduced early to astringent drinks, such as *ocha* (green tea), *kocha* (black tea), and *mugicha* (barley tea). Tea practices vary greatly. The Japanese are connoisseurs of green tea, but the Chinese sometimes use it to wash their bowls. The net effect of Asian food training is to develop familiarity along the neutral-to-sour axis of taste. The Japanese can tell Thai from American rice, and many prefer rice from specific provinces within Japan. Various noodles, from different regions and manufacturers, prepared in different styles, have Asian partisans as fierce as Italian pasta lovers. Traditionally the Japanese consume soup for breakfast: this *miso* is fish-based, slightly sour, and contains cubes of tofu or vegetables. Has globalization ended these practices?

No, but it has made *miso* a fast food available at Mister Donut. It has put *ocha* and *ramen* in vending machines. The persistence of the soup meal, in particular, can be seen among Asian students worldwide, who microwave instant *ramen* any time, anywhere. Noodles, ranging from *ramen* noodles to the whole-wheat *soba* and thicker *udon* were among the first solid foods they ate as infants.

Elsewhere in the world the basic starch comes from beans, yams, potatoes, cassava, or bananas, for other peoples build taste systems just as elaborate and enduring as those in Japan. One does not acquire a taste for marmite (Britain), fish sauce (Southeast Asia), *kimchi* (South Korea), or lichee nuts, turtles, and eels (China) except in infancy. Marmite is a foul-smelling, evil-looking yeast extract, which the manufacturer's own research shows that if an infant does not eat it by age three, the adult is unlikely to ever consume it. But 24 million jars a year are sold, and Marmite is exported to expatriates in thirty-three countries. As for eels, I saw children of four or five in the night markets of Nanning, China, who were already pulling them from the "live tanks" because they were so tasty. Nor as it turns out can we be reprogrammed to do something as simple as drink water while walking around, if we did not grow up doing it. In Italy, Nestlé has found it extraordinarily difficult to sell walk-around bottled water, despite the fact that each Italian drinks 189 liters of bottled water a year, the most in the world—the image-conscious Italians consider it rude and unsightly and they just won't do it. *Che maleducato!*

Is it surprising, then, that American fast-food restaurants have to modify their products extensively in order to sell them overseas? After all, Cantonese restaurants in the United States sell fat-laden, sweet-tasting "sweet and sour pork" that no one in Guangzhou would recognize. Tacos are served in the United States with a dollop of sour cream and grated cheese that make Mexicans laugh. American *sushi* chefs put not only cream cheese but *lettuce* in their products. Starbucks sells a lot of green tea *frappuccino* in Japan and Taiwan. The tastes learned in the high chair—oops! that's American baby furniture—are so important that all imported cuisines must lean toward them. In Japan, Italian restaurants are judged on the variety and quality of their noodles—spaghetti, linguini—and their sauces. These dishes never have the fat or morsels of meat, peppers, and tomatoes that Americans demand of "rich Italian sauces." The most popular fare in the French restaurants of Japan are Provençal dishes— fish, garlic, and vegetables—modified to Japanese taste—no *tapenade, patés* or *fromage de chèvre*. Cheese, especially strongly flavored cheese, has never caught on, although green-tea-flavored ice cream has. Häagen-Dazs is available but is treated as candy. The Portuguese introduced bread in the late 1600s, but the Japanese crossed it with the more familiar Chinese bun. The resulting *pan* is a white, spongy, eternally fresh product similar to America's Wonder Bread. The Japanese can eat their *miso* and *pan* breakfasts at many fast-food venues.

In France there are stronger seasonal variations in food than in the United States or Japan. Truffles and mushrooms appear in the fall. The first *radi* (radish) of spring is an event. The chestnut crop is reviewed on the radio. Was there *Beaujolais Nouveau* at your Thanksgiving dinner? Globalization has made November's *vin nouveau* into an international event, with thousands of crates sent by airfreight to Japan and the United States. The French *primeur* system, which prizes the earliest and the best of cheese, ham, wine, olive oil, mushrooms, truffles, beef, clams, apples, and beer, among other items, bears an uncanny resemblance to the Japanese fetishization of large, perfect fruits, the annual winter consumption of *fugu* (poisonous blowfish), and the appreciation of subtle differences in *tempura* or *uni* (sea urchin) and *unagi* (eels). The French also have distinctly different regional cuisines, some of them resolutely antimodern. Two types of potatoes from the 1800s were making a comeback when I last lived in Avignon.

Foreign cuisines falling in the midrange of the French taste system do not catch on. There is no appeal in German or Hungarian food. The French prefer not Mexican tacos, but American chili (lots of tomatoes and red peppers). They have virtually invented a Vietnamese dish they call *nem,* in which a deep-fried spring roll is wrapped in mint leaves and Romaine lettuce, then dipped in a sweet-sour sauce. Now *nem* exists in Vietnam, but it is very different. Nor can we find this version of *nem* outside France. Spain is a near neighbor to France, but little of its cuisine has caught on, perhaps because it is based on *arroz* (rice), potatoes, *chicharos* (chickpeas), and *alubia* (a bean and meat stew). Only Basque cooking has straddled the border. Morocco and Algeria are neighbors who have sent millions of migrants to France, but *cous-cous* and *kifta* remain immigrant specialties.

National-taste systems are reinforced by grocery stores. In every country that I have visited, "foreign foods" are rigorously segregated and they cost more. There is a food taxonomy called the "foreign foods aisle," where we are presumed to be either foreigners or looking for a foreign eating experience. Say a bottle of wine is our object. In a French Géant Casino, foreign beer and wine are stocked not with the general beer and wine, but in the foreign foods section. We must browse the respective shelves of the United States, Australia, and Chile. We find American wine cheek-by-jowl with peanut butter, Uncle Ben's rice, and corn flakes. Even jams, cookies, teas, and canned goods are arranged by nationality. On the Japanese shelf (next to the American shelf), Sapporo beer and Kikkoman soy sauce sit together with three kinds of instant *ramen*. They are the only Japanese products in this Géant Casino. The Italian shelves are well stocked, and there are a variety of Vietnamese products. For rice, there is Thai balsamati and Uncle Ben's, but no quality Japanese or American rice, such as Kokuho Rose. In the produce section, there are few imported vegetables, and those that appear have been given French names. Even Asian pears have a French name: *nachis*. There are no German or Spanish foods except some canned meats and olives. All coffees are French brands, most coming from former colonies in Africa. The teas are all packed by Tetley or Nestlé, and there is no real green tea, only *pamplemousse* flavored.

In Japan the segregation of foreign food (*yoshoku*) from native (*washoku*) is even stricter. The Co-op grocery in Nishinomiya sells no foreign vegetables except Chinese cabbages and only a few imported fruits, such as oranges and bananas. The only American foodstuffs are breakfast cereals, jams, peanut butter, pancake mix, and catsup; the only French, a few cheeses and wines in the gourmet section. In the big supermarkets, such as Daiei, we can find these products plus the ingredients of Tex-Mex cuisine, spaghetti sauces, and Pillsbury cake and instant mixes. There are a few British products (no Marmite), French Ma Mère preserves, various expresso coffees, and European chocolates. Budweiser, Gallo, and Guinness are for sale in the beverage section. This selection varies little, whether the supermarket is in the countryside or in Tokyo. Of course, if we disdain Daiei, we can find, at a price, anything we want in Tokyo.

Global merchants such as Unilever, Nestlé, Carrefour, and Procter & Gamble have found entering the overseas grocery business rough going. In Japan, consumers expected Carrefour to be "French." When they found it selling the same goods as Daiei at similar prices, they stopped coming. There were no deals on L'Oréal or cognac. In Hong Kong, Carrefour could not meet the local obsession with super-fresh produce and had to retreat. Nestlé does not sell *blé* in Japan but rice cakes, made in China, that undercut the price of Japanese-made rice cakes. In China, Carrefour now stocks a wide variety of *miantiao* (noodles) and soybean products, which China must import to meet the demand for soy.

Foreign food companies took a beating in China until they learned that the Chinese believe in "cooling" and "warming" foods, the yin and yang of cuisine. Lemon is a cooling flavor, so if we are Frito-Lay and want to sell potato chips in summer, we have to add a lemon flavor. But Frito-Lay (part of Pepsico) spent years trying to sell spicy flavored potato chips in the summer—duh! If Chinese want to cool off, they don't grab a soft drink—they eat lichee nuts. To warm them up in winter, KFC now sells a spinach, tomato, and egg soup.

The ways that people purchase and transport food also vary. Only in a few nations do people buy large quantities of food and drive it home. Even those French or British or Japanese who buy at supermarkets and transport by auto will purchase smaller quantities than Americans. At Géant Casino in France, I rarely saw purchases of more than twenty items, and at Daiei in Japan about eighteen to twenty items were usual. Refrigerators and storage space are limited. Almost no one has a deep freeze. People take their food home on foot, and they shop several times a week, if not daily. Some Japanese groceries deliver.

Magnus Pike, in a sober article on "The Influence of American Foods and Food Technology in Europe," wrote back in the early 1970s that "foods widely consumed in America have not been accepted in Europe and have made no impact there. . . . Only when an innovation from America fits, in some way that it is not possible to foresee in advance, into the social context of Europe, will that food or food process exert a significant influence there."[8] And foods are so basic to local culture that thirty years later that assessment is still valid.

Gender

Workers in the *maquiladoras* along the United States–Mexico border are predominantly female. When I asked why he hired only women, one Mexican manager responded "because they are better with their hands." It is tempting to focus on the sexual stereotype in his remarks, but they also reveal the stubborn perseverance of gender roles. Residents of the *maquiladora* zone are clearly globalized. They did not grow up on the border but migrated from poor, traditional states of the interior, such as Zacateca. Such migration is a feature of globalization, as is employment of women in light industry.

But globalization hasn't broken the primal patterns of gender. Gender identity, the individual's self-conception as male or female, is not fixed at birth, according to most scholarship. However, as the *Encyclopaedia Britannica* notes, "Basic gender identity—the concept 'I am a boy' or 'I am a girl'—is generally established by the time the child reaches the age of three and is extremely difficult to modify thereafter . . . gender identity develops by means of parental example, social reinforcement, and language. Parents teach sex-appropriate behavior to their children from an early age, and this behavior is reinforced as the child grows older and enters a wider social world. As the child acquires languages, he also learns very early the distinction between 'he' and 'she' and understands which pertains to him- or herself."[9] In Zacateca, women did all the work inside the house, while men worked outside. At the age of four or five, girls began to sew, cook, and clean. Boys went to the fields to learn about hoeing, animals, and irrigation. By age twelve boys might be working as cane cutters or coffee pickers, or clearing brush. Gender roles are so ingrained that when outside employment dries up in a Zacateca town, the *men* leave to find *outside* work elsewhere, even if there is local inside work available. This usually means they go north of the border. There are so many Zacatecan men in the United States that they have their own clubs and Internet pages. They send hundreds of thousands of dollars back every year in remittances, funding houses and clinics and schools back home. Globalization has simply made available a *different way* for them to fulfill a traditional gender role.

But there is a second class of Zacatecan families, headed up by single or deserted women, or poor and landless families who lack mobile male labor. Moving to the *maquiladora* zone, where the women become the principal breadwinners, represents a break with some traditions, but it retains the most basic gender role: women are good with their hands, and they work inside. As Hofstede notes,

> Men are *on average* taller and stronger, but many women are taller and stronger than quite a few men. Women have *on average* great finger dexterity and, for example, faster metabolism, which makes them recover faster from fatigue but some men also excel in these respects. . . . Every society recognizes many behaviors, not immediately related to procreation, as more suitable for females or more suitable for males; but which behaviors belong to which gender differs from one society to another. . . . The role pattern demonstrated by the father and mother (and possibly other family members) has a profound impact on the mental software of the small child who is programmed with it for life. Therefore it is not surprising that one of the dimensions of national value systems is related to gender role models offered by parents.[10]

Gender constructions remain stubbornly local. In the Chinese countryside, peasant women have worked outside planting rice paddies, even plowing with water buffalo, especially since the Cultural Revolution. They also work inside the house, and in the market and in shops. Men work outside and inside, as cooks, merchants, tailors, and metalworkers. Neither the Cultural Revolution of Mao Zedong nor the Great Leap into globalization have changed gender roles much. In the Guangxi province in 2003, I found a husband-and-wife blacksmith team, repairing an agricultural tool. She swung the heavy hammer, a task requiring some strength, while he turned the object and tapped the edges with a light hammer. Women also worked on construction sites, tended buffalo, and killed chickens in the market. Globalization didn't appear to change gender roles greatly.

But in Chinese cities and Taiwan, young women work in great numbers in manufacturing, which has led feminist critics to conclude that globalization has fomented a new form of "patriarchal enslavement." Although clearly sympathetic to these claims, scholar Aihwa Ong, in her exhaustive survey of the literature on gender and labor in Asia, ultimately arrives at the conclusion that "changes in the working daughter's status, with its mix of (and tension between) family obligations and growing personal autonomy, must modify sweeping assertions that pre-existing East and Southeast Asian 'patriarchy' alone is to blame for the construction of unequal industrial relations." Hong Kong daughters receive greater family support in return for their "filial" conduct: "As new workers," Ong writes, "young women engage in activities that violate traditional boundaries (spatial, economic, social, and political) in public life, forcing a redefinition of the social order." It's not "Fordist production" or "despotic regimes," she concludes, but "local milieux constituted by the unexpected conjunctions of labor relations and cultural systems, high-tech operations and indigenous values."[11]

In Iraq the *New York Times* found that feminists, including the Harvard-educated minister of public works Nasreen Barwari, supported polygamy and unequal inheritance laws. The paper pointed out that women are 55 percent of the population and that male-female relationships outside marriage are frowned upon, so becoming a second wife has practical advantages. As for inheritance, it is always men, not women, who are called upon to help their relatives, so they need resources.

Even in a more "American" situation, local traditions dictate gender roles. Thirty-year-old Madhauri Varik has an M.A. and works selling chemicals to biotechnology labs in Bombay. With her husband Gautam, she earns five times the national average. But when they attend Water Kingdom outside Bombay, she rents a bathing costume that covers her from neck to ankle—she doesn't own a bathing suit and shudders at the mere thought of a tank suit, much less a bikini. "Our culture doesn't allow that," she told the *New York Times*.[12] Park owner Ashok Goel says, "The swimming gear has been tailor-made for the Indian psyche. You can cover from the wrist to the ankle." The appearance of women at such a venue represents globalization, to be sure, but the women come in groups. And they all cover up.

In his essay "Algeria Unveiled," Marxist Frantz Fanon long ago pointed out the resistance of the local through clothing. The French colonists had mobilized "their most powerful and most varied resources," in Fanon's view, to remove women's veils and to westernize Arab society. In his hyperbolic account, the French were "committed to destroying the people's originality and under instructions to bring about the disintegration, at whatever cost . . . of the status of the Algerian woman."[13] In the ensuing battle (which the French lost), the colonized "displayed a surprising force of inertia." Were Fanon still alive, he might take satisfaction that fifty years later, neither globalization nor the French have made a dent in Arab gender roles, not even when Arabs resettle in France.

Western feminists cannot understand why Japanese women, in the world's second largest economy, do not push harder for an equal share. But Japanese feminists have responded, "Who wants those jobs—commuting two hours each way, working 10 hours, and attending business dinners night after night?" They point out that Japanese women control the purse, doling out allowances to their husbands, even buying and selling cars and houses without consulting their husbands. The situation of the Japanese "office lady," or OL, who is often college educated, has seemed particularly disgraceful. Japan's tabloid press noted that many OL's lived at home and spent their incomes on Gucci handbags and travel to Hawaii or Club Meds in Jamaica. The so-called Yellow Cab phenomenon (OL's who were sex tourists) was a national embarrassment. But as Karen Kelsky has demonstrated, there was less than met the eye. The traveling OL simply extended an unendearing Japanese custom of patronizing foreign sex workers. And the tradition of *mitsugu*—"giving financial aid to one's lover"—had been around since the fifteenth century. To glimpse the cultural essentialism at work here, we need only read *Beddotaimu Aizu* (Bedtime Eyes), by Eimi Yamada. Her protagonist explains that "[the foreigner's] smell seemed to assault me, like some filthy thing. But it also made me feel, by comparison, clean and pure. His smell made me feel superior."[14] In fact, all of the female sex tourists interviewed by Kelsky intended to wed Japanese men.

The gender roles of other cultures also baffle the big international retailers. The French luxury conglomerate LVMH took the success of its Vuitton handbags as a sign that its perfumes would succeed in Japan. It opened a flagship Sephora store in Ginza in 1999 and planned forty more nationwide by 2004. But its Chanel and Clinique brands attracted few buyers, and LVMH had to close the stores in 2001. "Japan is a skin care market," said a competitor, "not a fragrance market. The Japanese woman does not want to smell."[15] Gap and DKNY likewise found the popularity of their heavily logoed clothes among Japanese women short-lived. Instead, women flocked to the no-brand, post-logo simplicity of Japan's Uniqlo, whose simple, cheap outfits never go out of style. Japanese women "resent their bodies being used as advertising billboards," says President Tadashi Yanai.[16]

The most sensational topic in the gender and globalization debate is the sex trade in underdeveloped countries. There are scholarly books on sex tourism in Thailand and journalistic exposés on the Philippines and Indonesia, not to mention Michel Houellebecq's panegyrics. Their common theme is the guilt of the West (or flouting it, in Houellebecq's case). These critics are especially guilty of ignoring the local wellsprings of gender roles. In Thailand, the usual narrative is that Phuket and Pattaya were developed as R&R centers, with attendant prostitution, by the U.S. Army during the Vietnam War. But Thai prostitution is much older, as David

Lehney has shown: "The Bowring Treaty of 1855 . . . opened Thailand to foreign laborers. Most immigrants were young men from rural south China, planning to earn money for their families by mining tin in Phuket. A large number of Chinese prostitutes accompanied the men, establishing the largest sex centers Thailand had experienced at that time. . . . A 1909 law to prevent the spread of venereal disease effectively legalized prostitution."[17] Thai women ousted Chinese women from this lucrative business, manifesting a very different culture of communicative distance and an economic nationalism. Suffice it to say that many Asian cultures have traditions of bodily service. The blind are masseurs in some nations, while temple monks do this work in others. Sex work is paid well relative to other work, especially manual labor, which is always available. In Thailand women own stores and are managers, but even here sex work pays more, and some uneducated young women from the provinces enter it expressly to earn enough to return home and set themselves up in business. Others are runaways, "sold" by parents, or abducted by brothel owners. Without underestimating the viciousness of the sex trade, especially in eastern Europe and Africa, we should also see that economic self-interest and local gender constructions play a significant role. Women can always work for less money in tourist restaurants, in hotels, in boutiques, as guides or artisans.

Globalization did not create the sex trade, in short. Rather prostitution has been modernized by technology, just as food and clothing have. Air travel, tourism, media, and the Internet have changed the production and consumption of all three. There are more sex workers in the United States than in any foreign country, but they advertise in tabloids and on the Internet, and they operate through cell phones and pagers. They accept credit cards and schedule using Palm Pilots. As John Burdett writes, "the sex industry in Thailand is smaller per capita than in Taiwan, the Philippines or the United States."[18] But it's on the street and visible in Thailand.

We may feel that the spread of literacy, availability of education, improvements in health and communication and technology should make sex work evaporate. But in reality there is a tight embrace between gender, local culture, and a worker's return on effort that trumps our idealism. In the United States, a culture dominated by the Protestant ethic, educated women of marrying age patronize fitness clubs, but in Asia women take diet pills and sit in electrostimulation machines that zap them with a mild current. The fit bodies of young American women look too "working class" in Asia. For that matter, few Asian or European men seek the muscles (or avoirdupois) of their American brothers. This difference in body image is indicative of the links between local culture, work, and gender. Even in the effort to "look good," there are different local equations for effort and result.

So gender roles change as a result of globalization, but not essentially. In developed nations men are still steelworkers, repairmen, heavy-equipment operators, farmers, concrete and stone masons, yardmen, lumberjacks, and garbage collectors. Women compose most of the secretarial and clerical class, most of the elementary school teachers. Women are still the principal providers of child-care. What has changed is that educated women have accessed the professions in developed nations, becoming lawyers, doctors, engineers, architects, and professors. This change is most pronounced in the United States and Europe, but less visible in Japan, Taiwan, South Korea, or Singapore. Changes in this segment of the labor force may accompany globalization in some cultures, but are they particularly "American"? Inglehart says that "development of effective birth control technology, together with unprecedented prosperity and the welfare state, have eroded the functional basis of traditional norms."[19] Beyond this primary cause, he notes, the most influential secondary influence on gender roles has been the "historically Protestant" orientation of American culture, aligning it with Canada, Denmark, Sweden, and Switzerland. It is in these nations that gender roles have shifted somewhat.

With the dilution of the Protestant work ethic, though, things may not be changing at all. Between Ronald Inglehart's 1981 and 1990 surveys for *Modernization and Postmodernization,* almost every nation showed an increase in the number of people agreeing with the statement "a woman needs to have children to be fulfilled." The rate of agreement in countries such as the United States and Finland, which had been only 18 percent and 12 percent in 1981, increased to 20 percent and 19 percent in 1990. In France and Japan almost 80 percent of 1990 respondents agreed with the statement. These results contradicted Inglehart's prediction that postmaterialist societies would shift to a matrix of "self-satisfaction" factors. But some of the nations most dramatically globalized in recent decades, such as Hungary and South Africa, increased their rate of agreement with the women-children criterion by more than 10 percent between the surveys. Inglehart vouches only that "in sexual norms and gender roles, we find a continued movement away from the rigid norms that were a functional necessity in agrarian society."[20] Gender roles have been changed by modernity, certainly—by the presence of television and washing machines and cars. But globalization's impact seems to vary in accord with older, deeper gender templates, facilitating many cases of innovative adaptation and even some cases of return to traditional gender roles.

Education

Children see us working, or not, and begin to imitate us. We encourage them to read, to help with planting, to watch television, or to hang around the shop. By the time they arrive at kindergarten, *crèche,* or preschool, national educational systems are already different, as childhood historian Roberta Wollons has argued.

In Japan, there is a standard national curriculum, and few private schools. Teachers are highly trained, and families support education wholeheartedly. Most mothers stay home to raise children, some falling into the *kyoiku mama* syndrome—total involvement in the child's success or failure. By age three a majority of Japanese children are in preschool. As Joseph Tobin and Merry White have separately pointed out, the Japanese classroom appears surprisingly chaotic. Japanese teachers think their most powerful tool is children's view of them as benevolent providers. They maintain order indirectly, by encouraging the children themselves to deal with classmates' disorder. Tobin details the case of Hiroki, a boy who liked to pull out his penis in class. Waves of social pressure wore him down, from the little girls who scolded him to the boys who would not play with him to the principal who gave him a good talking to. Teachers encouraged other children to take responsibility for correcting Hiroki's behavior, but he never sat in the corner for a "time out," as he would in the United States. Exclusion from the group is too severe a punishment in Japan.

Among the assumptions of Japan's system is a profound egalitarianism. There are no gifted or slow students. There is no attention deficit disorder. Everything is a matter of applied effort. As Tobin writes, the Japanese have a "distaste for the notion of inborn abilities" because "the identification of children as having unequal abilities would inevitably lead to an unequal allocation of educational effort, resources, and opportunity."[21] "Intelligence" is associated with morality, obedience, and good behavior. Teachers did not think Hiroki was brilliant and bored. If he were more intelligent, they reasoned, he would behave better.

The Japanese educational system is the site of an enveloping conformity. In addition to the concepts just mentioned, its tools are ridicule, shunning, and bullying by other children. Any deviations from the norm are noteworthy, and the student's mother is the all-important coach. She must include exactly the right foods in the *bento* (lunch) or her child will be teased. As Susan Chira has shown, citizens of Korean descent learn not to send Korean foods. By five or six, Japanese children have fully absorbed this system; they ridicule the accents, clothes, and habits of returning students whose parents have been posted overseas. Students of the same sex, age, and school form extremely tight coteries, which remain important for the rest of their lives.

Ninety-nine percent of Japanese children attend elementary school, and 98 percent finish high school, even though attendance is required only to age fifteen. Children attend school 240 days a year, compared with 180 in the United States and 160 in France. Until May 2002, Saturday was a full school day. The change to a five-day school week was widely denounced, and all schools are still open on Sunday for clubs and sports. The involvement of Japanese mothers in PTAs is legendary, and they are often at the school and in the classroom helping out. They purchase an average of three books or magazines per month per child. As work becomes more complex, some mothers attend *mamajuku*, private schools that keep them ahead of their children so they can tutor them. Homework starts in first grade, and Japanese high schoolers spend more time on homework than students in any country except Taiwan, according to Thomas Rohlen. There is a core curriculum that focuses on language and math early, adding social studies, English, and the arts later. Everyone takes calculus and learns two musical instruments. Everyone takes gym, and a high proportion of students are involved in sports. Female involvement in sports is exceptional, comparable with that of the United States and Scandinavia. While 65 percent of American high school seniors spend less than five hours a week on homework, only 8 percent of Japanese seniors do.

The momentum achieved by this system dominates Japanese life. Since there are no "gifted" or "slow" individuals, everything depends on personal effort. "Seven times fall down, eight times get up," say the Japanese. Class assignments are usually made to groups, called *han*. Chosen by the teacher, the *han* includes a range of individuals and a *hancho*, or leader, whose job is to plan the work, encourage the slower members, and report to the class (this is the origin of the English "honcho"). Though individual talent is noticed, it is the value of a member's contribution to the *han*'s work that weighs most heavily. This system is carried to the lab, the office, even the factory floor, where all workers can read blueprints and perform higher math.

The American occupation after World War II affected the Japanese educational system only slightly. Five years of English are taught, in deference to the need for a lingua franca. But a powerful national teachers' union and involved parents have protected the system from four recessions in fifteen years. This system does not "reflect" Japanese culture; it *is* Japanese culture—local and impervious. When the European Economic Community commissioned a report on Japanese education—"how do they do it?"—the study concluded that Japan was a nation of willing workaholics, "masochistic," and willing to live in "rabbit hutches" without complaint.

Interestingly, the French system starts out similarly. Infants are coddled and rarely punished. The French ideal, increasingly rare, is the stay-at-home mother, but many children of three or four are taken to the *crèche* or *école maternelle,* usually a warm, colorful place that seems to enact the educational theories of Jean Piaget. My children attended elementary schools in the Vaucluse department, none of which had formal music, art, or physical education facilities. At my daughter's school, gym class consisted of running lengths of the town's soccer field. Art and music were taught in the main classroom; all students had to buy and to master the recorder. My son's school in Avignon had a separate *cantine,* where hot daily lunches were cooked by unionized culinary workers, but at my daughter's in the village of Lourmarin hot lunches were prepared by parent volunteers and served outside in good weather, or in the *rez chausée* in bad.

The French teacher reigns supreme in the classroom. Until 1998 parents had no legal right to enter their children's classrooms, unless requested by a teacher. When I attended the first parents' night that year, several people remarked that they had not been inside the school building since they attended classes there. Students were separated by sex for education until the mid-1970s. Schools are officially nondenominational but observe all Catholic holidays.

The French teacher does not orchestrate social conformity, however, but disciplines verbally. At the extreme she or he will call a student an *idiot,* an epithet likely to be repeated on the playground or to parents, and thus widely feared. Almost as bad are *bête* and *crétin.* Parents rarely use these words with their own children, instead doting on *sage* (wise), as in the often heard "Soi sage" (be good). But the teacher's authority to use harsh language is unchallenged. So too is his or her power to designate exceptional or slow students. France is, according to Hofstede, among the highest "power distance" societies in Europe, once run by royalty and since then by equally distant authority figures whose power one would never question. When I attended that parents' night, the teacher actually said, "At your homes, you rule. But in this room I rule." The parents nodded their heads in agreement.

Unlike the Japanese teacher, who may have several group activities going on at once, with parent helpers assisting, the French teacher presents a single subject to the whole class. When my son attended fourth grade in Avignon, his whole class studied the French Revolution and Jeanne d'Arc. They read difficult texts, some written in the *passé simple*. They were required to write in ink on block, not lined, paper. Math problems had to be solved in neat grids on lined paper in ink. Presentation was very important. Geometric drawing, using a compass, was another aspect of the fourth-grade curriculum; students were to render three-dimensional drawings of cubes, cylinders, and pyramids.

When French students enter *collège* (junior high school in the United States), studies become more difficult and an inegalitarian weeding out begins. The French do not recognize attention deficit disorder either, and they posit a degree of *génie* (genius) as the base of a child's abilities. The French are more disciplined than Americans about effort and practice, but *génie* is not something one develops. This view, combined with the end of mandatory attendance at fifteen, persuades many students, particularly from immigrant families, to drop out after *collège*. *Lycée* (high school) is quintessentially academic. It does not offer auto mechanics, home economics, ceramics, typing, or industrial arts. Newer *lycées* have physical education, but most female students contrive to avoid active participation. The lack of school sports for women is in striking contrast to the Japanese and American systems.

Those who continue on face the BAC, a national graduation test. The BAC is legendary, rivaling the Japanese university entrance exam. A nationwide, government-sponsored exam, the BAC takes place over several days and covers many subjects. Today there are several versions of the BAC, but they retain the commonality of being more philosophical and more dependent on written expression, even written neatness, than exams in the United States or Japan. Students headed for the same profession will answer the same questions in Brest, Strasbourg, Marseilles, or even in Dakar. Years later, people who took the exam in the same year find themselves discussing their answers. The BAC is not only the capstone of the French secondary education system, but a unifying national experience, so successful that versions have been exported to other European nations, to the underdeveloped world (where the French *lycée* system is sometimes the only real education), and to the United States, under the rubric "international BAC."

In the schoolyard, French education is also different. French students do not wear uniforms like the Japanese. They aim for the casual look, but there are no sweatpants, and the beauty standard is high. There is more racial mixing than in the United States: students of "Magreb" origin who make it to *lycée* are more socially accepted. Asians are also well integrated, Africans less so. Teachers do not generally interfere in fights or disputes, even though the schoolyard is a domain of petty theft, lying, and *tricher* (cheating). This reinforces a traditional French distinction between public, institutional behavior and private behavior.

There are few sports at *lycée*, and nothing remotely resembling Japanese baseball or American football for school spirit. Rather France's gifted soccer, basketball, swimming, and track stars all develop through private clubs. Because athletics is early disconnected from the high school scene, the "jock" does not really exist. Events like the *randonnée*, the school field trip, become more central than "the big game." In good weather, teachers take students on walking visits to museums or historic sites. In Avignon, my son went to the Palais des Papes. In the village of Lourmarin my daughter's trips were walks in the countryside during which local plants such as *tilleul* and *menthe* were gathered. This breaks up the strict age group cohorts that exist in the Japanese and American systems, for there is constant interaction across age groups. It also reinforces the weekend practice of *se ballader,* or walking about, and the intense intra-France tourism of the French.

The French school year contrasts strikingly to the American and Japanese. Students attend school all day Monday, Tuesday, Thursday, and Friday, and half days on Wednesday and Saturday. The French explain that the half days are necessary so that children can recover from the intensity of the two previous days; officially, the Catholic Church is supposedly providing religious instruction on Wednesday afternoons. After the third grade the school day begins at 8 a.m. and ends at 5 p.m. However the lunch or recess break is two hours, and there are usually two thirty-minute or longer breaks, so the school day is only six hours long, like the American one. School begins around September 6, and there are long vacations for Toussaint, Christmas, Easter, and numerous saints' days and national holidays. The net effect of this calendar is a rhythm of intense effort followed by frequent vacations. This calendar meshes with the French life-style of frequent short vacations to *la campagne* or *la mer.*

The idea of months upon months of uninterrupted hard work is utterly foreign.

Despite quirks and dissent, the French education system is enormously popular, and it has been exported all over the world. French students not only *do well* in foreign universities but in the world job market. French engineering, business, and medicine rank with the best. Globalization has not changed or even nicked the quality or reputation of the French education system. And it has in no way "Americanized."

Work

One would expect work patterns to change drastically because of globalization. As far as point-of-sale scanners and cash machines go, we'll see in chapter 3 that work has changed. But *ways* of working and *attitudes* toward work have proved remarkably resilient. In separate surveys of thousands of subjects worldwide, Inglehart and Hofstede reveal just how culturally embedded "work" remains. "Scarcity has prevailed throughout most of history," writes Inglehart, but then "industrial society developed the belief that scarcity could be alleviated by individual achievement and economic growth."[22] New value systems arose, with shifts in attitudes toward "power distance" and "uncertainty avoidance."

> The root *cause* of the Postmodern value shift has been the gradual withering away of value systems that emerged under conditions of scarcity, and the spread of security values among a growing segment of the publics of these societies. This, in turn, grows out of the unprecedented high levels of subjective well-being that characterize the publics of advanced industrial society, as compared with those of earlier societies. In advanced industrial societies, most people take survival for granted. Precisely *because* they take it for granted, they are not aware of how profoundly this supposition shapes their worldview.[23]

But even as scarcity abates, the attitudes that it shaped endure, as Hofstede showed (see fig. 2). In general the "small power distance," "weak uncertainty avoidance" cultures have prospered (Denmark, Sweden, Ireland, Britain) and the "large power distance," "strong uncertainty avoidance" cultures have stagnated (Panama, Greece, Guatemala).

In both Inglehart's and Hofstede's emplotments, Japan and France share some of the latter qualities, but they have done well. Their stan-

TABLE 4

Hours Worked in the Industrialized World

Nation	Annual Hours Worked	Weekly Hours Worked	Number of 40-Hour Weeks / Year
United States	1,979	38	49.5
Japan	1,842	35	46.0
Canada	1,767	34	44.0
Britain	1,719	33	43.0
Germany	1,573	28	39.3
France	1,556	27	39.0

SOURCE: Steven Greenhouse, "U.S. Growth Industry: Workdays," *International Herald Tribune,* September 3, 2001.

dards of living are close to that of the United States, making a comparison of their work habits interesting.

As shown in table 4, Americans lead the industrialized world in average annual hours worked, at 1,979 hours. The Japanese are second at 1,842 annual hours. The French are farther down. On average Americans worked three and a half more weeks than the legendary Japanese in 2000. We might suppose that if vacation and holidays were added to the right-hand column, the figure would total fifty-two. Two and a half weeks of vacation and holidays sounds right for Americans. But do Germans have nearly thirteen weeks of leisure? The Germans take four weeks in the summer, plus a week at Christmas and another at Easter. With assorted minor holidays (workers can take off their birthdays in many cultures), the German total is closer to seven weeks, like the French. Another difference is that government-imposed work weeks of thirty-five to thirty-eight hours, designed to spread the available work around and especially to create entry level jobs for young people, strictly limit the hours spent in shop, office, or factory. Attempts to increase the hours worked per week, due to globalization, were among the reasons the French voted against the EU constitution in 2005, according to analysts.

The U.S. worker is more productive, generating $54,870 in 2000 (in constant 1990 dollars). The Belgian worker was second at $53,370. However, if we look at productivity per hour worked, the French worker leads ($33.71) followed by the Belgian ($32.98) and then the American ($32.84). The French worker is 2.6 percent more productive per hour. This is not the impression of Americans visiting Paris and waiting for café service, of course. But if we think back on the *lycée* system, with its weekly rhythm of intense work followed by rest, its relegation of sports and extracurricular activities to the side, its mechanical drawing, and the fear of being labeled an *idiot*, we get some clues. The French education system required conformity to French standards of presentation, style, and hierarchy, but that simplifies work life—the result is greater productivity per hour with more free time. However there are side effects. There is not much part-time work available to youth, unemployment is often high (10–12%), and good jobs are hard to find. But when people find them, they tend to stay, so that average job tenure is much longer than in the United States. Thus less training has to go on, and employees know their work in greater detail.

What about the legendary Japanese? In Japan, part-time jobs for young people are plentiful; every high school and college student seems to have one. Most of these jobs are in retail sales, food, and service, but other students tutor or give music, language, *ikebana*, or computer lessons. None of these jobs has benefits, and most produce only minor income. Yet the Japanese hold in high regard students who perform well in *arbeito*. Countless aphorisms tell the benefits of early hard work, even failure, so that one appreciates later success. Many aspects of these jobs are highly ritualistic, such as always arriving on time, addressing customers in the appropriate *keigo*, providing fast service, knowing details of the items for sale, and standing motionlessly at a work station when unoccupied. To watch these young cashiers, waiters, or gas station employees during a peak period is to witness total focus.

The intensity of this experience prepares the young Japanese for the experience of finding and holding an adult job. Government statistics placed the jobless rate at 5 percent in 2001, when I lived in Japan, but that was meaningless, since part-timers were counted as "employed," while the underclass of *kojiki* (beggars) and *homoresu no hito* (homeless people) were not counted at all. Figured on American principles, unemployment in 2001 was probably 10 to 12 percent. Obtaining a real job is difficult and time consuming, a process that college students begin in their junior year. They change their hair, their clothes, and their habits, disappearing on two- and three-day interview trips to distant cities.

The first real job is a momentous event, comparable with the university entrance exams or marriage. Although all Japanese know that lifetime employment is *not* guaranteed today, the enormous investment in finding the job and the structure of the job itself make staying with it the logical thing to do. Most people remain at their first jobs for fifteen to twenty years. In Japan I knew a young man, Kenji, who had just found a place in a bank. His boss chose a residence for him, which was a dorm where he roomed with three other new bank employees. He rose at 6 a.m. to study banking for an hour before work. He arrived at work by train at 8 a.m. He had no desk. All day long he stood at the beck and call of senior employees, who assigned him to empty ashtrays, to retrieve files from the basement, or to check long lists of figures. He worked until 6 p.m. Two nights a week he practiced with the company soccer team for two hours. Other nights he was required to go out eating and drinking with his seniors, often until 11 p.m. or midnight. At dinner he was required to sing songs and to explain company policies, and ridiculed if he made mistakes. After two years, he would become a junior-grade official with a desk, joining the upward march on the salary scale, which is rigidly geared to age and years on the job. Having made such an investment, he would be loath to change jobs.

When I visited his bank, workers sat at desks side by side and facing one another, more than fifty in one room, without partitions or anything between them except computer screens. Cubicles would have been luxurious. Personal effects were limited, the work space was cramped, and customs—who makes the *ocha,* who greets visitors, who opens or closes the office, even who goes first through a door—were rigidly observed. Progress through the ranks, the minor perks that come with it, are very visible to a work force that ranges from twenty-one-year-olds to retirees of sixty or sixty-five. This behavior is highly *inefficient,* and Japan's banks, unlike its auto companies, are not known for their streamlining. Like society, the bank was highly structured, and my young friend felt that he *must* reproduce that structure. Service to the company became almost a filial obligation, and he even adopted the archaic habit of his elders of referring to the bank as *san,* the honorific suffix for humans.

In Japan productivity at work is achieved through a culturally reproduced atmosphere of group-think and artificial scarcity. Japanese banks are not limited by space; they could have large offices. They don't need to maintain written files in the basement; everything could be on line. They don't need dorms, soccer teams, or nightly dinners. "Organisms," Richard Lewontin has written, "in the course of their evolution as a species do not *adapt* to environments; they construct them. They are not simply *objects* of the laws of nature, altering themselves to the inevitable, but active *subjects* transforming nature according to its laws."[24] The culture of Japanese banking explains Japan's ATMs, which are open only from 7 a.m. to 11 p.m.—that's when there are suitable things to buy.

In contrast to the Japanese, according to both Inglehart and Hofstede, the Americans and the French are "postmodern" in their work culture, but in different ways. Take AOL America and AOL France. According to the *New York Times*, AOL America epitomizes geek chic, from the work attire of sandals, ponytails, tie-dyed T-shirts, and backpacks to the desks covered with Legos and Star Wars figurines. AOL's American programmers eat Szechwan, sashimi, and microwaved burritos at work, are used to a very low power distance, tend to be purists (Unix, C++, Apple), and expect not only complete freedom to innovate, but that their inventions will drive the entire business and that they will be rewarded generously. They will work twenty-eight hours straight, or knock off after four hours to go surf.

But at AOL France the programmers wear suits or sport coats. They start at 8 a.m. and leave exactly at 6 p.m. They all take the same lunch hour and eat at the same table in the cafeteria. Jacques Sireude, technical product manager for AOL France, says, "The job is a job. We are here to work. You are not at home."[25] Earlier Sireude worked for a software firm in Santa Barbara, California, where he went body boarding when he finished a project. "It was very cool," he said. "Developers were at the same level as product managers—even the director." But he moved back to France because he preferred French social customs, the emphasis on more and longer vacations, and the quality of life outside work. According to Steven Riou, an AOL manager in Europe, the work habits of software developers are different in each nation. Fine business suits, good haircuts, and first impressions are the norm in Italy. In Germany "major technical decisions will be taken over by marketing people—even if we know the limitations of the problem. They just want it to work." German software developers feel less connection between their work and the company's direction or profits, he said. British software developers were very reserved, rarely asking questions, and typically lifting only a finger when they wished to speak.

Land Use

Local land use has been surprisingly resistant to globalization. The persistence of historic agrarian patterns, when farming is in decline and cities are growing, is impractical and irrational. But land use patterns are anchored not only in economic stages of development or physical needs but in cultural ideologies.

The American suburban model of land use has been thoroughly documented by urban historians. It developed out of independent, individual transportation, beginning with the horse. Then streetcars and automobiles transported the middle classes to the ideal detached house, surrounded by green lawn, in a neighborhood close to urban amenities but free of noise, crime, and traffic. Attempts to change this, to create denser cities and mass transit, have mostly failed, as the *New York Times* documented in a multipart series in May–June 2005. The attractions of suburbanization have dictated more economic opportunity, with "better" schools and cultural amenities thrown in as a justification. Accompanying this shift has always been the assumption of independent transportation through cars and highways. Such land use is in fact a huge brake on globalization, which strives for efficiency and flow. Suburbanization is striated with redundancy, such as high per house costs for water, sewer, electric, and roads—not to mention environmental problems. Every new subdivision retards the efficiency, focus, and speed that characterize globalization. Average transportation spending as a percentage of U.S. household spending was 19.2 percent in 2001–2, with areas like Tampa topping 23 percent. So ingrained was this habit that the huge increase in gas prices during 2004–5 did not affect it.

JAPAN

The resistance posed by local land use is even more obvious in Japan. The stereotype of Japan as a small, densely populated island is only partially true. South Korea, the Netherlands, and India are more densely populated. At 337 people per square kilometer, Japan's population density is comparable with that of Israel. Mountains and history, though, have restricted the Japanese to living in certain areas, but unlike the Israelis, the Japanese have created the illusion of scarcity, especially in land, when in fact vast areas of Hokkaido and northern Honshu are unpeopled. But there are rice fields in Osaka, where the population reaches 4,000 people per square kilometer. How this happened is instructive.

In the 1700s and 1800s the Japanese shoguns controlled their nobles by requiring them to live at court every other year. The latter had to reside in strictly demarcated zones close to the *jo* or castle. Their retainers lived just beyond, the merchants just beyond them, and so on. Beyond all were the farmers. Everyone's residence was restricted and permission was required to move. Back home the nobles controlled huge tracts of land by levying a "rent" equal to 48 percent of the farmer's production, creating a class of serfs who survived only by borrowing from unscrupulous moneylenders. Shoguns and nobles set aside some mountains as hunting preserves or for timber, and gave others away to Buddhist or Shinto temples. Good agricultural land was relatively limited, mostly in river valleys or deltas. This centuries-old pattern was still in place when World War II began. Sixty-six percent of the countryside was still worked by tenant farmers paying the 48 percent rent.

After World War II, American occupation forces insisted on immediate land reform. New laws established a limit on individual holdings, permitted the seizure and resale of large holdings, and gave tenants right of first refusal as well as a long payment schedule. A 1947 follow-on law decentralized rural government, so that village affairs were decided by villagers. "Within two years, tenancy declined by more than 80 percent," writes one observer. "Rent control and land distribution helped to equalize incomes in the villages and rehabilitate the sociopolitical status of the peasants."[26]

But there land reform froze. The Liberal Democratic Party (LDP), which was still in power in 2005, wooed farmers in the 1950s and has represented their interests ever since. The revenge of the peasantry, modeled on the shoguns, is now codified in law. The new landowners hold their property with fanatic intensity. "Despite improved techniques, the output per worker [has] declined," writes James Brooke. "Low per capita income, underemployment, and insufficient mechanization have persisted. Even black market rents developed."[27]

How could there be black market rents when farming does not make money? Because the LDP subsidizes all agricultural production, pushing rice tariffs as high as 490 percent. The *New York Times* called it "an economic fantasyland in which a farmer can earn $50,000 a year from three acres."[28] But that sum doesn't go far in Japan, especially when pesticides and fertilizers and machines (or hired labor) are involved. Even with subsidies, the number of farmers fell from 12 million in 1960 to 3.9 million in 1980 to 2.8 million in 2004. The remaining farmers are mostly older than fifty, and their children have fled to cities. But the land use pattern continues. Millions of acres of former farmland are covered by scrub forest, and wild boar have become a problem in abandoned rice paddies. But farmers continue to hold their land as a "savings account," because land has always been so valuable. Their small farms are now like a tourniquet around some urban areas. As figures at Demographia.com reveal, in the boom period of 1985 to 1995 the agricultural use of land in Japan's top three urban areas (Tokyo, Osaka, Nagoya) declined only from 13.5 to 12.3 percent. There was a net gain of just 0.5 percent for urban residential land. Just 25 percent of the whole country is designated as "city planning areas," but 114 million people (92% of the nation) live on that land.

As a result, urban land is very expensive, or perceived to be. People who own urban land never sell it either; instead they rent or lease it out. As land has increased in value, both rural and urban landowners pressed the LDP to pass laws enforcing their right to be free of high taxes, zoning restrictions, and eminent domain. Today we can see farmers in major cities, often in traditional garb, tending their urban *tambo*, which will produce only a few hundred dollars worth of rice even with government subventions.

In the aftermath of the 1995 Kobe earthquake, tens of thousands of Japanese were displaced, their homes destroyed. To the north, the large relatively fertile valley of the Mukogawa River looked like a logical rebuilding area. Its largest city, Nishinomiya, was where I lived. Every building on my block was new, all built on the former *tambo* of a family who used government subventions after the earthquake to change their vocation. They *leased out* most of the land for construction of single-family houses, using the cash flow to pay the construction loan on an apartment building that provided their income (and my residence). They made the leap from peasant farmers to real-estate magnates, living in a large, traditional house of several thousand square feet at the corner, keeping an eye on everything. Checking the local land records, I found that the extended family of my landlord—fourteen to eighteen families with the same name—owned most of the land in the *cho*, but it was broken up into very small parcels. It would have been impossible for Daiei or Wal-Mart, for example, to buy enough land from one of them to build a store: it would need two or three, who would have to swap land, which would entail negotiations within the extended family, who were all watching out for their future property values. It took a devastating earthquake to convince these families that the time was ripe.

This land use pattern produces, albeit indirectly, the legendary apartments called 1DK (a bedroom plus a dining room-kitchen), where an individual and sometimes two live in twenty to forty square meters. Life in this space must be extremely organized and precise. The building materials are of low quality: doors are thin sheets of metal, windows do not close tightly, walls are uninsulated and thin; their coverings, moldings, and even the floors are thin plastic simulacra. Maintaining quiet is a rule, and loud music or entertaining are violations. Kitchens are small, refrigerators tiny. These apartments are shoddy because the building will be torn down in twenty-five years, the owner hopes, and something else built.

Public space is treated just as niggardly. As Alex Kerr noted in *Lost Japan* (1996) and *Dogs and Demons* (2001), Japanese cityscapes are appallingly ugly. Urban land use planning has yet to be discovered, and *no one has thought about how the city looks.* James Fallows also commented that "how difficult and materially constrained work and residential life were for the average Japanese person, during a time when everything you read about the Japanese economy emphasized its steady rise to greatness."[29] *Touché.* In my brand-new postearthquake neighborhood in 2000, city planners had a chance to do everything right, but they replicated existing land use patterns. Houses stood hip to hip, and there were no sidewalks. A spaghetti of electrical and telephone lines crisscrossed the sky, hanging on telephone poles placed several feet out into the streets, frequently reducing them from two lanes to one. Electrical transformers hung and hummed in public view, and storm sewers were open culverts at the edge of the pavement. About once a week a tow truck arrived to extricate a motorist gone over the edge. There were no street signs, and not a single public parking space, despite every home housing at least one auto. Where it connected with the major cross street, my street narrowed from four lanes to two and then to one. But there were rice *tambos* and vegetable fields every few blocks. The use of land was irrational, but it was traditional.

Japanese railroads also dictate land use. The world's quality king in auto making discourages people from driving. Because of sacrosanct land rights, freeways must be built *above* cities, at stupendous cost. Interstate freeways go *through* the mountains, at equal expense, to avoid precious farmland (and to support construction companies). Because cities laid out on shogunate policies have tiny streets and nowhere to park, most people need to live within walking distance of a commuter railroad; these are the largest railroads in the world, as we saw, and their wealth enforces the existing land use pattern. In fact, the railroads have extensive surface holdings and easements. They own large department stores. Every railroad station creates a neighborhood, and everyone wants to live within walking or bicycling distance of a station.

Even with a car, finding destinations is difficult. Visitors need to ask directions repeatedly. Satellite navigation systems in cars were an instant hit. The majority of streets are one way, compounding the problems of getting lost. There is no free public parking, so parking is scarce and expensive. For an urban apartment, just parking a car may cost an extra $400 a month. My landlord charged $200 a month for a space. Department stores offer free home delivery, even of groceries, but most people transport what they buy. Housewives shop for food daily or several times a week. Package sizes must be smaller for convenient carrying. Detergents and laundry soaps are concentrated, in boxes with carrying handles. Land use dictates everything.

Interestingly, the *homuresu no hito* (homeless people) live by exploiting the contradictions of traditional land use patterns. Their blue-tarped boxes appear along beautiful riverbanks (designated as floodplains, so no one can live there), in large city parks, and under railroad bridges. Not only do they often live farther apart than the middle classes, but with great views of nature. Most of them earn enough to eat by collecting cans and glass, which they sell to recyclers. They also reuse clothes, appliances, televisions, and bicycles, which the Japanese throw out every few years, so they diminish the landfill problem. Land use traditions are so strong in Japan that the occupants of this liminal zone now have established vested rights too.

MEXICO

In Mexico the indigenous peoples have held land communally at least since the founding of Mexico City by the Aztecs around 1300 B.C.E. and probably longer. Colonization by the Spanish, who imposed the *hacienda* system and city plans laid out on the boulevard-and-plaza template of Spain, did not replace indigenous land use—rather the latter eventually became codified in *ejidos* and *communidades*. These have proved remarkably resistant to globalization.

For several centuries, the *haciendas* and Catholic Church attempted to break the land use patterns and keep *indigenos* in serfdom. Then Benito Juárez pushed through pro-peasant land reform in 1865, but the terms allowed *haciendas* to scoop up as much church land as they lost to the peasants. The Revolution of 1910, in the view of most scholars, was a delayed reaction to the unintended consequences of Juárez reforms. Emiliano Zapata, coming from Chiapas, where the peasant system was strongest, tried to undo the government's earlier seizure of lands. But as scholars Yetman, Burquez, and Womack have detailed, land reforms have focused on rural, mostly Indian, villages, where lands had been illegally absorbed by neighboring *haciendas*. However, the constitution of 1917 affirmed the rights of *all* property owners, including *haciendas,* and contained enough loopholes to complicate and perpetuate the existing system. The *haciendas* remained, as did a form of traditional collective land-holding called *ejidos.*

Ejidos were collectively owned but individually worked lands, held in trust by the Mexican Ministry of Agrarian Reform. "Actual decisions as to the uses and ultimate disposition of the land were to be decided by the *ejidatarios* in mandatory monthly meetings of the *ejido* elected assembly," writes Yetman.[30] *Ejidos* emerged from their fuzzy legality under the presidency of Lázaro Cárdenas (1934–40), when millions of hectares were redistributed to peasants. By the end of the 1940s great advances were transforming farming in the developed world, such as crop rotation, nitrogen-fixing crops, erosion prevention, pesticides, and hybrid seeds. But 69 percent of all Mexican farmers were *ejidatarios* working a few hectares at best, and their per capita output continued to fall. Some of the new agricultural science appeared on the *haciendas,* which were increasingly turned to cropland. Thus by 1950 an oligarchy of one-half of 1 percent of all farmers owned 31 percent of the actual productive land in Mexico. During this period Mexico's population increased dramatically, rising to fifty-one people per square kilometer by 1990 (by comparison, U.S. population density was thirty-one per kilometer in 2004). Basic foodstuffs such as rice, corn, and flour had to be imported. The need for more agriculture production was pressing, but reform proved impossible due to the *ejidatarios.* Although some later presidents opposed the *ejido* system, they made no dent in it. Presidents José López Portillo (1976–82) and Luis Echeverria Álvarez (1970–76) actually expanded the *ejido* system to paper over political mistakes. Yetman writes that by the early 1990s, more than 230 million acres (one-third of Mexico, and one-half its arable land) were in *ejidos* controlled by 3 million *ejidatarios,* less than 3 percent of the 2000 population.

Divided into small parcels, separated by hedges, walls, and easements, remote from or lacking rights to water, *ejidos* offer only a meager living. Yet they are embraced not only by the peasants who nominally "farm" them but also by political patronage. The *ejido* bank (Banco Nacional de Crédito Rural) has been "a virtual bottomless pit of disappearing public funds," writes Yetman, requiring no collateral, ignoring repayment, and dispensing bribes and favors so prolifically that it could service only 40 percent of eligible *ejidatarios.*[31] Presidential candidates used the bank to write "production loans" to farmers before elections, warning them that all subventions would disappear unless they voted correctly.

Meanwhile Mexico imported more and more food. Some of its old *haciendas,* incorporated as agribusinesses, found exporting fresh vegetables to the United States more profitable than growing for Mexicans. *Ejidos* were too small for mechanization and lacked water, so they were used for grazing. Irrigation is capital intensive and intrinsically more efficient on large landholdings, where there can be fewer canals, pumping stations, and "faucets," more direct routing, and less evaporation. What was required was agribusiness. But foreigners were forbidden to own Mexican land (until recently), so agribusiness stayed away, and anyway the property rights of foreigners in Mexico have had a history of disappearing.

By 1990 *ejidatarios* were not engaged in some vestige of noble subsistence farming; many were absentee landlords, working in U.S. cities or leasing their land to others. So many *ejidatarios* leased out their land illegally that entire *ejidos* existed without a single resident. *Ejidos* are a prime example of local land use resisting globalization, even to the extreme detriment of the local population. Land reforms in the 1990s should have given *ejidatarios* full ownership. But in Sonora, according to Yetman, very few *ejidatarios* have applied for full ownership. One who has is considered a "land baron" by fellow *ejidatarios* because he assembled two hundred hectares, enough to run thirty head of cattle. He nets $10,000 to $12,000 a year. A cattle operation of this size is not commercially feasible across the border in Arizona or Texas, which *are* the competition in the global market.

This situation is not likely to change soon, because "a twenty-hectare parcel of eroded and overgrazed land far from infrastructure and without water is hardly a juicy plum for either speculator or empire-builder," as Yetman notes. His findings in Sonora "indicate that displaced or unemployed rural Sonorans tend to migrate to Sonoran cities rather than directly to the United States."[32] Contrary to public perception, the Mexican migration problem is not *caused* by globalization, but rooted in land use patterns that globalization has not been able to shake.

The *Economist* suggested recently that the real problem is not NAFTA

but Mexico's failure to adapt to trade liberalization in general since the mid-1980s, when it first acceded to the General Agreement on Tariffs and Trade (GATT). Since gaining access to all those shiny new markets in America and the European Union, Mexican agricultural production has either declined, collapsed, or grown only slightly. For all types of beans, for instance, production fell on average by 0.7 percent a year between 1980 and 2001. Wheat production has fallen by 57 percent since 1980, and soybean production by about one-sixth. NAFTA merely accelerated all this. Mexicans, and world markets, have preferred cheaper alternatives.

Mexican governments failed to take advantage of the ten-year transition period, while the tariffs were being phased out, to invest in infrastructure improvements such as irrigation. It is the high cost of Mexican farming that makes it so uncompetitive . . . local farmers are still going out of business because their costs—from diesel to electricity to credit—are about a third higher than those north of the border. Poor transportation makes a crucial difference: it costs about three times as much to deliver corn by rail from Sinaloa to Mexico City as it does to ship it there from New Orleans via Veracruz.[33]

In the summer of 2002 these farmers, citing their land rights, took two dozen hostages, including the state prosecutor and police officers, in efforts to stop Mexico City from building a new airport eighteen miles west of the city. The land in question already boasted Mexico City's largest dump, but peasants didn't want to sell their farming rights (in a nearby swampy lake bed) for the market price of about one dollar a square meter. They stole and burned thirty vehicles and blockaded the highways into the capital as well. Ironically the "farmland" they were protecting used to be under Lake Texcoco, a reservoir in the Aztec water system. Environmentalists say its most important use would be as waterfowl habitat.

FRANCE

In central Avignon all roofs are red tile, and all walls are stone or brick, most covered with stucco. The building codes specify these materials, which are regionally produced. Residents like the look. There are many houses from the 1700s, which the city government deems "historic," and their exteriors cannot be changed. Some of them are uninhabitable shells: no wiring or plumbing, no windows or wallboard. One of these, near my apartment, was a "historic" stable, whose owner negotiated with the city for five years for permits to renovate it as an apartment. But the neighbors agreed that it was a "charming" ruin and, furthermore, that since it had always been a stable, it should remain one. Actually they didn't want the neighborhood gentrified, since a higher *taxe d'habitation* would result on their houses.

If the exteriors of buildings in Avignon are locked into the architectural template of 1700, their interiors are quite different. The French have cut away walls and floors, installed gleaming kitchens, and designed appliances and furniture that fit antique walls. From the efficient *chauffe-eux* to the *lucarne* (skylight), this modernization fuels a French design industry and Mr. Bricolage Home Centers. The French can't do anything to the exterior, but they can make the interior a warm cocoon.

But then the French dichotomized inside and outside centuries ago. The outside of a house may be shabby—to deceive the tax assessor, to underplay the occupants' wealth—while the interior is lavish. High walls, hedges, and fences forbid the public view of private life so common in American suburbs. The French have collected art, amassed fortunes, committed murder, and dreamed of *madeleines* behind walls that needed a coat of *crépi*. This *inside* draws value from its contrast to the *outside*. As Hofstede and Inglehart note, the French exhibit a large power distance, strong uncertainty avoidance, low belief in God, but a rather individualistic nature, and since 1947 they have voted along class lines more consistently than other Western democracies.

With the French *commune* locked into ancient land uses, new development has been relegated to the periphery. Freeways called *periphériques* go around the city, because of the sacrosanct remnants of city walls. Gilles Blieck has shown how these walls, originally built to protect local lords or the church (and often worthless defensively within a few years), were early on the subject of debate between Renaissance "preservationists" and French progressives who wanted to rationalize street planning and urban life. Preservationists rallied the *commune* by asserting that any change in the walls (which had been modified within fifty years of their construction) was an attack on the property rights of the people. By 1900 the walls were not only untouchable but in need of constant maintenance, and today cities such as Avignon, Carcassone, and Roussillon spend millions of Euros on wall repairs.

Out beyond them are broad, modern highways lined with Castoramas, Géants, Auchans, and McDonald's, the French versions of the United States strip mall. Mixed among these are the high-rise public housing developments of the poor and immigrants. Impoverished versions of Le Corbusier's vision of "machines for living," these invite neglect by their lack of connection to the core of the *commune*.

The "right" of the *communard* is to have things remain exactly the same: to have a neighborhood bakery, gardens, and places to stroll, and streets with interesting vistas. The French are connoisseurs of the city scene. "Ruining" an urban vista is a charge that will humiliate an architect. Buildings must be carefully placed in the landscape or cityscape, a tradition older than Louis XIV's gardens or Haussman's boulevards. Doisneau, Atget, and other French photographers have memorialized the *commune*'s vision of a continuously inhabited and historic common space. This is reinforced by the educational system, not only in history classes but in the visual arts. With those rulers, compasses, and block paper that my son used, French students learn to view objects, buildings, and scenes in three-dimensional space. They learn to rotate those scenes and objects in space. Texts for beginning architecture students emphasize not bridges or buildings, but imagining an urban renovation from multiple points of view. How is the street altered by strategically planted trees, new pedestrian zones, or the elimination of on-street parking? The student must imagine change from multiple points of view before a decision to "improve" is acted upon. This results in a cityscape that has what Grant McCracken has termed "patina," or emotional validation.[34] It explains why the city permits the Croq' Monsieur fast-food chain to open but denies McDonald's. Though the former will produce more litter, hire fewer employees, and produce less tax revenue, it will preserve the building façade and have a modest sign, in French. It will not jar the eye with its golden logo or insist on stamping its brand over three hundred years of communal conformity. Such an ethos applies widely in France. If the streets are too small for large foreign cars, well, the three French auto makers produce excellent small ones. If apartments lack elevators and their refrigerators are small, well, the French shop several times a week—so stores should be closer, and products smaller and lighter. Pay attention to how things *look* too: building exteriors may be stuccoed brown, but the toaster and toilet needn't be ugly.

Unlike Japanese farmers, French farmers are prevented from profiting on the increased value of their land. They are not holding it until the irresistible offer comes along, because the *commune,* the *département,* and the province decide where and when development can occur. That happens only when everyone with a right has had a say. The collective moves slowly. It has a long memory. If new development were allowed, what new roads, or expensive bridges would be needed? What noise or inconvenience would be created? Whose business would be harmed by competition? What historic monuments be threatened?

The *commune*'s wheels are sometimes difficult to grease. When Paris explored sites for a third airport in 2001, it found a promising one to the north, but a World War I cemetery for foreigners occupied some of the land. French reporters went to England, seeking someone whose rights might be violated. They found a 103-year-old British survivor of the war and presented his halting objections to millions of television viewers. It was enough to derail plans for the airport site.

The most notable failure of French land use has not been transportation, as one might expect. France is covered by excellent roads that connect the smallest hamlets. Rail systems and airports are top-notch. The "failures" are those block apartment buildings outside the city. There the former colonial subjects of Algeria and Cameroon live, loathing their dwellings. Immigrants are not allowed to infiltrate the center of the *commune,* as they do in New York or London (an aid in assimilation, as Jane Jacobs long ago pointed out). Instead a majority of France's 5 million Muslims live in low-income, high-rise buildings called *HLM* (*habitation loyer moyen*), which have nurtured auto theft, drug traffic, and cells of al-Qaeda. Police are often afraid to enter these buildings.

Meanwhile the *communes* refuse to expand beyond their ancient borders, much less raise taxes to provide new infrastructure and schools. North American–style housing developments are a rarity. A French woman can commute a distance of fifty kilometers from a village to her job in a suburban *grande surface* in thirty-five to forty minutes. As a result, villages along the excellent *autoroutes* and close to train stations are packed. Parisians too keep weekend homes in villages. Having one foot in the country (or pretending to) while working in a city or the capital is an old French tradition, the reason why Paris has more small hotels than any city in Europe. When I lived there in the 1980s, my landlady housed another paying lodger in her garden shed! When it comes to land use, the French have decided to pretend that they are all still *villageois.*

Tribalism

In the fall of 2001, an American baseball player named Tuffy Rhodes was closing in on Japan's home run record. He was only two homers away on September 14, but most Japanese had never heard of him. He was never mentioned in the Japanese press, received no product endorsements, and he had no press conferences or fan mail. He was not in a single TV commercial or on any talk show. He was black. This was going on while Ichiro Suzuki was garnering Rookie of the Year awards in the United States. Baseball is one of the few places where *nihonjinron,* an avatar of tribalism, breaks the surface in modern Japan. *Nihonjinron* means the superiority of Japanese culture, but it implies the superiority of the Japanese race.

Another black American player, Leron Lee, had ended his career with the highest batting average in Japanese history a decade earlier. But as Robert Whiting notes, "Team posters for the Lotte games frequently omitted Lee's name in order to stress the Japanese-ness of the team. There were autograph parties for the Japanese stars on the team, but never for Lee." Japanese executives said that "Japanese do not want black role models."[35] Still earlier, in the 1970s, Randy Bass, a white American slugger for the Hanshin Tigers, mounted an assault on Sadaharu Oh's home run record and got up to fifty-four. But Oh was the manager of the Yomiuri

Giants, the opposing team in the last three games. "Rather than risk letting their manager's record fall," wrote Howard French, "the Giants' pitchers tried to intentionally walk Bass during every at bat over three games. The American player managed a single base hit only because he threw his bat at a pitch."[36]

Back to Tuffy Rhodes. He tied Oh's record of fifty-five on September 24, 2001, and with five games to go, appeared to have the record in hand. However, his team played its last two games against the Daiei Hawks, managed by none other than Sadaharu Oh. The Hawks walked Rhodes in every at bat.

"Tribalism" is a leaky roof under which to gather aspects of racism, nationalism, and sentiments of ethnic or cultural uniqueness. By highlighting its power, I am not excusing it. Research has shown that racial differences are genetically minimal. We know that poor peoples were once powerful, that culture's "essence" is as much imagined as real. The emptiness of these categories is beyond dispute, but tribal sentiments are still there. Intrinsically local, building on gender, food, language, and land use, tribalism resists globalization. Inglehart and Kecmanovic, whom I have cited earlier, have shown that tribalistic forces are stronger determinants than religion in creating personal identity. People of the same race and religion but different tribes have been on opposing sides in wars, for example, in Afghanistan and Iraq. It is equally a mistake to locate tribal impulses in the underdeveloped world, as Benjamin Barber does in *Jihad vs. McWorld* (1995). Japan is the world's second largest economy, the world's largest donor of foreign aid, and a constitutionally pacifist country. Tribalism is everywhere—in the United States, in Russia, in Mexico, in France—and globalization may be intensifying it by bringing tribes into closer contact, making their members more conscious of difference. Tribalism has survived laws penalizing it, regimes that outlawed it, churches condemning it, and billions spent educating against it. But even good parents unconsciously pass it on, and their offspring learn it in the street, at school, from popular culture, at work, or in the military. Leave your hometown and you experience it in a dozen forms.

Japan is no better or worse than any nation. It has laws against racial discrimination, and its schools preach multiculturalism. Per capita the Japanese travel abroad more than citizens of any other nation (15% of all adults in 2000), and more Japanese students go on foreign home-stays than students from any Asian nation. Yet the Japanese sense of themselves as a distinct tribe is astonishingly strong. It is in part fostered by the educational system that fetishizes conformity. As I noted in the education section, the Korean-Japanese student whose parents have been in Japan for a generation, who was born in Japan, speaks Japanese, and whose name is Japanese may be ostracized for eating Korean foods in his lunch.

Foreigners visiting Japan find that natives ask about their blood types almost immediately (most Japanese have one of two blood types). Those around Japanese children soon hear themselves called *gaijin*, a word with negative implications. "Look mom, it's a foreigner" is a phrase I heard regularly in Nishinomiya. But I heard Chinese, Koreans, and even Okinawans labeled similarly, for children learn as infants to identify what is "Japanese" in human appearance. Japanese adults pride themselves on being able to review a crowd of Asians and pick out the Japanese, in a way that an American reviewing a crowd of Americans would consider ridiculous. The Japanese are shocked when they go to China and hear themselves called *wo kou* (little bandit), an epithet sometimes applied to Americans and other non-Asians. And when a Japanese friend of mine visited Australia and was mistaken by local blokes for a Bushman, he was devastated for months.

The Japanese are deeply invested in a conception that *their* culture is unique. This is a feature of most tribes. Despite evidence of common genetics and of common religion, agricultural and building techniques, and linguistic borrowings, the Japanese insist theirs is a separate civilization. Archaeological discoveries purporting to prove the antiquity of prehistoric "Japanese" gravesites, or the early use of tools or techniques, are featured on the television news. Sometimes this is outright fraud, as in the 2001 case in which a leading academic researcher was shown to have salted archaeological digs with artifacts from other sites. Another scholar's assertion in 2001 that Admiral Perry threatened to attack Japan turned out to be based on a nonexistent document. But Japanese schoolbooks in 2003 continued to reprint both "truths," while omitting the Rape of Nanking in 1937–38 or Japanese atrocities during World War II. School texts that perpetuate tribal identity are also the norm, as we will see in the case of Israel.

Tribalism extends to the arts. Liza Dalby, who grew up in Japan and speaks fluent Japanese, has written in her book *Geisha* (1998) about the difficulties placed in her way when she set out to learn to play the *shamizen*. No foreigner has ever appeared in *noh* or *kabuki* theater, not for lack of language or talent but for lack of "cultural background." No foreigner can ever gain enough standing as a chef of Japanese cuisine to open a restaurant in Japan, despite the fact that the cuisine is no more difficult than French or Italian (many would say it's easier). But there are many Japanese chefs preparing expensive French and American food in Paris and New York.

Most of us, especially academics, regard tribalism as basically a pernicious mischief. In our culture, it has proved relatively amenable to education. As Senegal's president Abdoulaye Wade pointed out in 2001, racism against Africans in places such as Europe is "marginal" compared with ethnic conflicts in Africa itself. The underdeveloped world is far more racist, he said, pointing to Ivory Coast, where a wave of violence was occurring against immigrants from Burkina Faso. "A Burkinabe in Ivory Coast is treated in a way that a black person would not be treated in Europe," he added.[37] As if to prove his point, when word of his comments got back to Abidjan, Ivorians started looting and burning Senegalese shops.

Even as Wade spoke, thousands of Africans were dying in tribal conflicts in Rwanda, in Burundi, and in the Congo, where 3.3 million people were killed in battles between the Hem and Lendus ethnic groups between 1999 and 2003. Neighboring Angola's largely tribal civil war between 1975 and 2002 left 400,000 dead. Hundreds of thousands of Nigerians were dying in clashes between the Hausa and Yoruba tribes. The government of Mali, controlled by the Songhai and other black tribes, was fighting the Tuareg, who are a Berber and Arab mix. After a civil war that left 800,000 dead, Mozambicans were attacking immigrants from Zimbabwe, while in the latter President Robert Mugabe made the few remaining white citizens the object of his campaign of hate and expropriation. Eritreans and Ethiopians were locked in tribal struggle to the north (at least 30,000 dead), and the nearby Arab government of Sudan continued a ten-year-long war of extermination against the Bantus of southern Sudan (50,000 dead), who were darker than the northerners and regarded as de facto slaves. They were being exterminated at such a rate that they were granted special immigrant status by the United States. Thousands were brought to Boston, Texas, and Ohio, as detailed in the 2004 film *Lost Boys of Sudan*. *Atlantic* writer Jeffrey Taylor, in *Angry Wind* (2005), has detailed his year-long journey across the Sahara's southern edge, where every tribe he met was trying to avenge perceived slights from every other tribe in the area.

There are complicated contexts to these African conflicts, and European colonization may have played a role in consolidating them, as some scholars claim. But the opposing inference, that the tribes would have been peaceful without colonization, is demonstrably false. Despite apologists, despite colonizers, despite modernity, these are tribal wars. They have resisted the cures of development wealth, mediation, and UN peace-keepers. None other than Kofi Annan has pointed out that the great resistance to globalization comes from what he prefers to call "populism" and "nationalism" that "exploits historic enmities and foments trans-border conflict."[38] In most regions tribalism undercuts globalization, just as it undercut the colonial powers who, when they departed, left functioning governments and economic infrastructure. The colonial powers introduced sophisticated racisms, were hypocritical, and played favorites, but they did not create *tribes.* If anything, their net effect was to weaken them. Even the most empathetic reportage on the Rwandan genocides, such as Gil Courtemanche's *A Sunday at the Pool in Kigali* (2003), finds no out-side "cause" to tribal hatreds. The UN War Crimes Tribunal held in Tanzania found that Hutu newspaper publishers, radio broadcasters, and even pop star Simon Bikindi all urged the slaughter of Tutsis, which is to say that the technologies of globalization can be taken over to promote tribalism.

Most reporting, however, ignores the deep roots of tribalism. The *New Yorker* in 2003 published an article suggesting that the "gangsta" attire of rebels in Liberia and Ivory Coast was evidence of American role model-ing, as if T-shirts caused genocide. Ironically, V. S. Naipaul had predicted twenty years earlier in the same magazine that tribalism would undo the promising synthesis achieved by the first president, Houphuet-Boigny. The self-blame of Western liberals does not explain the African genocides.

Nor does blaming TNCs. Amid the controversy about Shell and Chev-ronTexaco in Nigeria, only the *New York Times* got to the tribal basis: "Two local tribes, the Ijaw and the Itsekiri [are] locked in a violent struggle over who should control the local government and the millions of dollars in patronage—and returned oil profits—that accrue from such dominion." Says a retired Nigerian judge, "Maybe if there had been no oil, these tensions would not have been created."[39] Or maybe if there had been no television or movies. No antimalaria drugs or dentistry. Imagining a pre-lapsarian innocence is no answer to tribalism. Annan refers to such apologists as "those who would return to imagined communities of earlier times."[40]

In the Middle East, besides the Arabs versus Jews, there are the Turks versus the Kurds, the Iraqis versus the Iranians (or Sunni versus Shiite), the Moroccans versus the Berbers and the Algerians (and the Spanish); Lebanon is a stew of primordial tribal groups, as is Egypt. A 1999 study by Daniel Bar-Tal of 124 textbooks used in Israeli elementary and sec-ondary schools found that Palestinians and Arabs were depicted as "murderers," "rioters," "suspicious," and "generally backward and unproductive." These texts used by 1 million Israeli Arabs—20 percent of the population—"are written by and issued from the Israeli Ministry of Education."[41] There are no Arab-language universities, and no Arab has ever won a major university scholarship in Israel.

One of the interesting sidelights of the Iraq conflict was the renaissance of the Kurdish nation. Threatening the "modern" nations of Turkey, Syria, and Iraq, the Kurds were apparently opposed by everyone except the United States. But they had one powerful if covert supporter. According to Seymour Hersh, Israel spent millions in the region and planned to use a Kurdish state as an advance barrier against Iran. The Iraq conflict also highlighted the tribal basis of Saudi Arabia to the west, where Abd al-Rahman al-Hariri told a reporter: "People here don't have a political consciousness of a nation. It's completely tribal underneath." Said another Saudi: "Our tribes are wilder than those of Iraq. Now our women want us to get guns too." The *Wall Street Journal* concluded that the Saudi tribes were "well-organized, tightly knit and well armed, and the urban middle classes and business elite fear that the old tribalism will fill the current power vacuum."[42]

The Basque separatist movement ETA in Spain plans to allow only "pure Basques" who are free of any "taint . . . of Spanish, Jewish or Arab blood" into its utopian state. One of the stranger parts of its dogma is the Basque language, which was spoken by only a few thousand rural inhabitants when Basque nationalist Sabino de Arana y Goiri revived it. Today no one can hold a public job in the Basque provinces without passing a test in Basque, though almost none of their grandparents spoke it.

To the south, mobs of Spaniards chased and beat Moroccan farm workers in 2000. The Moroccans worked in the 10,000 greenhouses along the Costa del Sol, at $25-a-day jobs that Spaniards rejected. We should recall that Spain was once colonized by the Moors and owes to Arab civilization many of its tourist attractions. Yes, Spain is "modern," but is also highly homogeneous, with a population less than 2 percent foreign born. Like other tribes, the Spanish blame immigrants for petty crime and rising rents and trash on the streets. So Spanish mobs destroyed Moroccan teahouses, two mosques, and then fire-bombed Moroccan shanties. The government waited three days to send in riot police, and Spanish labor unions refused to represent the greenhouse workers. "They came with sticks and steel bars," said Kamal Rajmuni, a community organizer. "People had to run for their lives."[43]

Two million Roma, or gypsies, live in the EU. The campaigns against them are usually led by local media and governments in the name of anticrime, antipeddler, or anticamping sentiment. Thirty years of pervasive and brutal discrimination against them have led the European Commission to establish an inquiry into anti-Roma actions.

Austria has a long history of tribalism, going back to the Middle Ages, and not only against Jews. When I taught at the Universität Wien in 1993–94, student bathrooms were regularly spray-painted with the graffiti "RAUS AUS" (foreigners out). There were similar signs outside Beethoven's house and on trail markers in the Wiener Wald. Croatians, Serbs, and Bosnians fleeing the Balkan War were the object of daily scorn by the *Neue Kronen Zeitung* and its favorite politician, Jorge Haider, who lived on four thousand acres expropriated from Jews in the 1930s. When Germany's Siemens tried to take over a failing Austrian engineering company in 2004, and thereby save a thousand local jobs, it was rebuffed. When Springer Verlag, a German publisher, tried in 2004 to take over Britain's *Daily Telegraph*, it met a blitzkrieg of German-hatred, including one publisher's statement that all Germans were Nazis.

Asia is also tribal. Even as India develops a software industry and high-tech call centers, Hindus and Muslims burn each other's temples and villages. In March 2002 Hindus on a train in northern India attacked Muslims and refused to pay Muslim vendors for snacks. The latter telephoned stations ahead, where their allies fire-bombed the Hindu cars, killing fifty-eight people. In retaliation Hindus across the state of Gujarat rioted, killing five hundred more people. As Katharine Boo revealed in the *New Yorker*, at those outsourced call centers in places such as Chennai "Brahmin and other upper castes dominate the supervisory class, while untouchables still clean toilets. They're lucky to have jobs at all."[44] Some Indians who have lived in the United States for decades still will not interact with or hire Dalits, as untouchables are called, much less intermarry with them.

In Thailand I learned that the natives look down on Cambodians to the east. In Cambodia, I found that Thais were resented, but the real tribal enemies were the Vietnamese, who work in the lowest trades, such as prostitution. In Malaysia, the Malay majority enforces government discrimination against the centuries-old Chinese of Penang and Malacca, as well as the Indian workers on the tea plantations. The *bumiputra*, or "sons of the soil," enjoy special legal privileges denied to both other groups. Each of Indonesia's thousand islands is home to a relatively unique tribe. Residents of Java and Sumatra live in different worlds, as do the Balinese (Hindus) and small Christian minorities in Maluku, Poso,

Mataram, and Medan. But all of them discriminate against the Chinese, who have resided in Indonesia for more than two hundred years. My friends in Java feel the need to "explain" Chinese blood in their families. The Pacific nations of Tonga and Fiji long agitated for independence, but after they won it and mismanaged their resources, they blamed "white outsiders." The king of Tonga blamed "whites" when his investments declined (in the worldwide downturn in 2000–2!). According to Subramani, one of Jameson's collaborators, the culture of Fiji is under attack by "*vulagi* Indians" who attempt to fit in through "a pastiche of unassimilated Fijian rites, for example *kava* drinking, pieties and fantasies from Bombay cinema, and thought and ideas from Western Education."[45] So much for academic and Marxist tolerance.

Tribal tolerance is no better in the Americas, even leaving aside the well-known problems of the United States. From the Mexican border south, tribalisms prevail that are largely invisible to visitors. They are based on "pure Spanish blood" and the degree to which this is evident in hair, skin color, features, and name. No politician of *indio* (Indian) ethnicity has ever been elected to a governor's office in Mexico, much less to the presidency. There is a great deal of intrafamily racism, with fair-complexioned children favored and darker ones neglected. In Honduras and Nicaragua, *mestizos* in government refuse to aid Indian farmers in the highlands. The governments of nations such as Ecuador and Peru make minimal pretense of guaranteeing Indian rights: Indian villages lack the sanitary water and sewage systems, schools, and hospitals that other citizens take for granted. At the better hotels of Quito and Lima, *indigenos* and *negros* were refused entry when I traveled there. Fifty years of communist rule in Cuba have not made its society one whit less tribal, according to testimony from Carlos Eire's *Waiting for Snow in Havana* (2004).

In Brazil, there were as many as 25,000 "forced workers"—slaves—laboring on the vast frontiers of the Amazon in 2003. They felled mahogany trees for market and burned brush; they worked with toxic chemicals without masks or gloves; and they moved and slaughtered the vast herds of free-range cattle. The government's antislavery strike force, which began in 1995, freed 288 slaves its first year, 583 in 2000, and more than 1,400 in 2002. "Particularly troublesome workers," wrote a *New York Times* reporter, "especially those who kept asking for their pay, were sometimes simply killed."[46] Need I add that they were dark-skinned? At the time this story broke, however, Brazilians were in a lather about an episode of *The Simpsons* in which Homer gets kidnapped by a taxi driver.

Any account of globalization that fails to consider the momentum of tribalism is missing a key component. Globalization exacerbates tribalism in three important ways. First, tribes are threatened by corporations, especially TNCs, which can replace them as supreme economic arbiters. As competition for resources heats up, a TNC may underpay, but the tribe kills. Second, globalization forces nation-states to specialize in what they do well, be it new technology, low-cost labor, or raw resources. This is the nation's "marginal advantage." Thus underdeveloped nations find themselves with only one or two items to offer in the international marketplace. Not having corporate or legal infrastructure, they default to tribal battles over the resources. Third, globalization encourages economic migration, thus placing tribes once isolated from one another in contact and economic competition. The proper practice of "politics," as Annan has argued, may assuage such conflicts, but there are no immediate examples. On the other hand, as Jeffrey Sachs points out in *The End of Poverty* (2005), the poorest nations are often the most landlocked (and tribal) and need nothing so much as transportation infrastructure that links them to the global supply chain—modernity, in other words.

Corruption

I crossed the outback of Cambodia in a pickup truck in 2000. Once an hour we were stopped, sometimes at a full, cross-road barricade, sometimes by a lone man in fatigues. We always stopped at bridges because planks of the roadway had been removed and lay in the hands of waiting locals, often children. Out came our bribe money and off we went. We even paid a bribe to drive through a river, since this involved crossing someone's pasture. All bribes had to be paid in foreign currency.

Phnom Penh was the same. There were meters in cabs, but that wasn't the price you paid. Boat or airplane tickets—you paid one price, the ticket said something else. When I registered at the local police station at 3 p.m. it was deserted, because the cops got freebies in the massage parlor across the street in the afternoon. In karaoke parlors I was offered handguns for sale, and my motorcycle driver wanted to sell me opium.

Cambodia has more excuses than most countries. The most recent civil war ended only in 1998, and most of the judges, lawyers, policemen, teachers, and government workers were slaughtered in the killing fields. The nation's infrastructure was destroyed, people in the outback were starving and afflicted with AIDS, and vast areas were without electricity or sanitary water. People did anything to survive.

Unfortunately, corruption has a gene for transmission and enormous momentum. In the underdeveloped world, systems of corruption initiate young people into the world of work. If you are a cab driver, there are police who will pull you over and take your driver's license, your plates, even tow your car away, unless you show them "a little respect." If you run a store, there are building inspectors who will shutter your business or turn off the power, unless their palms are greased. There are tax collectors who will approve your books, if you understand that they too have to pay off higher ups. If you want to buy property, build an addition, bring in a power line, or form a corporation, it will take years, unless you put up a bribe. John Burdett has written better than most about Thailand's thralldom to the bribe in his novel *Bangkok 8* (2003).

When Thomas Friedman was a reporter in Iran, it did not permit U.S. credit cards, so he carried cash. He had $3,300 when leaving. He describes the Iranian customs agent who suggested he fork over a $300 bribe if he wanted to keep the rest of his cash: "The two of us pretended to be rummaging through my open suitcase, and with a quick snatch of his hand he grabbed the $300 out of my fingers. It happened so fast—like a trout going for a fly—that you would have needed slow-motion instant replay to see it. Then, with his other hand, he handed me a new, blank customs form, which he asked me to fill out, declaring that I was taking only $500 out of the country."[47] Friedman called such nations "kleptocracies," but after Enron that sounds ethnocentric. Corruption is ordinary. In Iran, Albania, Nigeria, Egypt, Indonesia, and Russia, what we call "corruption" is an organized form of governance. There are two important aspects of this with regard to globalization. The first is that corruption is not "bad," since it is just an alternate economic form. But it is highly inefficient, arbitrary, and inequitable. The second is that it is a de facto form of resistance to globalization.

Corruption can organize entire societies. After the collapse of the Soviet Union, the communist government of Albania was essentially replaced by a political Ponzi pyramid. The leaders demanded payoffs from the politicians, the politicians extorted businessmen, and the businessmen extorted customers. Not surprisingly, the best extortionists were the Mafia, who soon controlled the country. Then "the Ponzi schemes began," said Carlos Elbirt, head of the World Bank office in Tirana, "with efforts to raise cash to finance the purchase of gasoline that could be smuggled at very high prices into neighboring Montenegro and Serbia, which were under international sanctions during the Balkan War."[48] Soon there were Ponzi schemes everywhere. They promised 20, 30, even 50 percent in six months. The largest scheme sponsored an Italian race car team. Even

Albanian employees of the World Bank could not resist, despite warnings from their boss. "People sold their homes and put the money in Ponzi schemes and then in two or three months they bought back their old home and a new one," said Elbirt. In 1997 the pyramid collapsed, resulting in a complete breakdown of law and order, with Albanians attacking state buildings and stealing anything portable. As he left the country, Elbirt's car was taken. The thief even demanded the papers to make it official.

Corruption is not the result of lack of resources. Nigeria is abundantly endowed with natural assets, including extensive oil and natural gas. It has been for more than two decades one of the most corrupt nations in the world. One of the few TNCs remaining there is Royal Dutch Shell, which operates oil wells off the coast and provides 80 percent of Nigeria's national revenue. According to Stephen Farris of *Fortune*, Nigeria pocketed $15 on every $20 barrel of oil produced in 2001. Production cost $4, leaving Shell with $1. That gave Nigeria $12 million a day, or $4.4 billion a year in oil revenues. But the government gave nothing to the people, particularly in the delta region where the oil is pumped and pipelined. In 2000 there were 176 attacks on Shell personnel, whom the government would not defend. Shell, on top of its royalty payments, was forced to make "community development efforts" that cost millions each year. It built schools, medical facilities, and wells, and it gave operating funds to the local tribal elite, who commonly prefix their names with "King," "Prince," and "Royal Highness." Initially Shell made annual disbursements to these "village leaders," but none of the money reached the village people. Now Shell doles the funds out monthly. But the schools are still empty, the clinics have no equipment, and the wells lack pumps. Almost every "donation" has been stolen. Shell has undertaken 408 such development projects, but by its own internal review only one-third can be counted at all successful.

Everyone in the field knows that huge amounts of international development money are squandered, but no one knows how to stop it. To develop local infrastructure, local people have to make decisions, to supply services, and to be paid. World Bank funds helped to develop the town of Dahab on the Red Sea in Egypt as a scuba diving and resort center in the late 1990s. Egyptians decided on the building loans, the contractors, and the diving operators. When I visited in 2003, half of the town was abandoned hotels, restaurant foundations, and curio shops. At peak capacity, the area might support twelve diving centers, but the government had licensed more than thirty to operate. Sewage ran right into the ocean that was the primary attraction, and the water supply went on and off. But there was a new airport, vast boulevards, and a new army barracks (to defend the resort). Anything that could be built of concrete or asphalt had been contracted out to the best-connected Egyptians.

So brazen has large-scale organized corruption become that some governments portray themselves as dogged by unfair international creditors when the credit stops. Nobody cried like Argentina, which defaulted on $88 *billion* in bonds in 2001. It wanted us to believe that only the IMF and perhaps Citibank suffered when it offered its investors twenty-five cents on the dollar. But 44 percent of its bondholders were small foreign investors, including 450,000 Italians, 150,000 Germans, and 35,000 Japanese. In previous sovereign default cases, settlements averaged 64 percent, but Argentines are uniquely self-regarding when it comes to paying up, as we will see in the tax discussion that follows. While it was attempting to leave foreigners with its debt, Argentina's economy grew at a 4 percent rate, and the government raised pensions, unemployment benefits, and government salaries. In 2005 it tried to reduce its payments to bondholders even more.

A variety of NGOs, academics, statisticians, and accounting firms, as well as the IMF, World Bank, and UN, have begun to publish annual reviews, based in some cases on more than 100,000 interviews, on definable aspects of corruption. Appropriately weighted and factored, these are now combined in an annual, composite Corruption Perceptions Index. Although what is measured is "perceptions," there are as many as twelve surveys for each of the hundred countries rated. The average is seven. At the top in terms of honesty:

1. Finland	9.9
2. Denmark	9.5
3. New Zealand	9.4
4. Iceland, Singapore	9.2
6. Sweden	9.0

But even the squeaky clean Finns were found in 2004 to have paid "commissions" of $9 million to the president of Costa Rica in order to sell the country medical equipment.

The United States ranked below the United Kingdom and Hong Kong in 2002, with four of the world's largest economies.

16. Israel, United States	7.6
18. Chile, Ireland	7.5

20. Germany	7.4
21. Japan	7.1
22. Spain	7.0
23. France	6.7

A full floor below come the first nations of South America, Africa, and the former USSR.

26. Botswana	6.0
27. Taiwan	5.9
28. Estonia	5.6
29. Italy	5.5
30. Namibia	5.4
31. Hungary, Tunisia	5.3
34. Slovenia	5.2
35. Uruguay	5.1
36. Malaysia	5.0

Below this point—note that Malaysia's score is half of Finland's—the bottom drops out.

42. South Korea	4.2
44. Poland	4.1
50. Colombia	3.8
51. Mexico	3.7
57. China	3.5
59. Ghana	3.4
61. Thailand	3.2
65. Zimbabwe	2.9
71. India	2.7
79. Russia	2.3
84. Kenya	2.0
88. Indonesia	1.9
90. Nigeria	1.0
91. Bangladesh	0.4

The bottom group is not subjective: Russia and Indonesia are covered by ten and twelve studies each. Cambodia didn't make the 2002 list: it was considered a convalescing economy. But in 2004 the World Bank stated that "Four fifths of the private sector sampled acknowledges the necessity of paying bribes, and 71 percent of large firms report that these payments are frequent."[49]

Also valuable is the Opacity Index created by PricewaterhouseCoopers. It measures the "clarity" of the economic environment: Are there uniform accounting standards and practices? Is there a uniform and honest legal system? Are regulations clear, and honestly enforced? Obviously of interest to corporate clients, these practices also reveal a nation's "threshold" of corruption. Only the thirty-five largest foreign investment markets are surveyed. Among them, China has the worst score, receiving the only 100s (i.e., totally opaque) for the impenetrability of its legal system and regulatory regime. It is followed by Russia, Indonesia, and South Korea. The last is commonly regarded as an industrialized nation, indeed one of the "Asian Tigers," but it gets the world's worst score for accounting. With its obscure legal system and lack of clear economic policies, Japan's overall "opacity" score placed it with Colombia, Brazil, Argentina, Pakistan, and Peru for investment climate. Better by far was Mexico, which had a higher corruption index, but clearer economic policies and regulations. It shares North American accounting standards. Singapore had the highest opacity score (29), followed by the United States (36) and the United Kingdom (38).

Many companies from the developed world contribute to corruption by offering, or at least paying, bribes. The "Bribe Payers Index" covers the nineteen leading exporters, who account for more than 60 percent of exports to non-OECD countries. They have all signed and ratified the OECD's Anti-Bribery Conventions. Sweden has the lowest bribery rating, followed by Australia and Canada. The United States and most EU nations rank next, except for more bribe-inclined France, Spain, and Italy. The French are rather regularly caught making bribes. In Costa Rica, it was revealed in 2005 that phone giant Alcatel had paid José María Figueres, president from 1994 to 1998, about $1 million in "consulting fees," then turned around and paid his successor, Miguel Ángel Rodríguez, $2.4 million from 1998 to 2002.

Considerably below this group comes Japan (14), which has signed and ratified the treaty. At the very bottom are South Korea, Taiwan, and China; the last two have neither signed nor ratified the treaty. For these nations, bribing customers is normal.

The types of projects for which underdeveloped nations say that bribes are offered is revealing. Public works contracts and construction lead the way. The building of roads, bridges, airports, harbors, sewers, and water systems (in which France and Japan lead the world) is the largest category of international aid. The second most bribe-prone sector is the defense industry (where the United States and EU lead), followed by the energy industry (oil, gas, electrical stations), and heavy industry (including mining). The least bribable sectors of the economy are agriculture, banking and finance, and civilian aviation. These surveys also show that small companies are twice as likely as TNCs to be asked for bribes. Almost 10 percent of them reported paying sums equal to 10 percent of their revenues in bribes, but only 3.7 percent of TNCs broke that barrier. For most companies, bribes were a relatively small cost. Only 14 percent of large firms and 33 percent of small firms reported paying more than 2 percent of gross revenues as bribes. Fifty-eight percent of large firms reported paying no bribes at all. Clearly, TNCs are less likely to tolerate corruption.

To whom do companies pay bribes? That depends on the pay level of the government employee and the anticorruption programs in place. A top priority of the World Bank, IMF, and UN has been funding initiatives in both areas. Extensive studies on the "Amount Corruption Adds to Bureaucrats' Salaries" have been carried out in Africa. Where government employees are poorly paid, such as in Kenya and Nigeria, bribes must be offered across the board. Employees expect an extra 50 to 100 percent of their salaries in bribes. Uganda and Eritrea follow these nations, with bureaucrats expecting an extra 10 to 50 percent income from bribes. Botswana, Namibia, and Tunisia have raised government salaries and started anticorruption programs: in these nations no bureaucrats are perceived to expect "tips and bribes." Aside from this trio, however, the proportion of bribes as a part of total government employee income has been steadily rising for twenty years across Africa. Companies report that in Nigeria most bureaucrats expected to double their official salaries in 1998.

A former French foreign-service officer described to me how bribery works at the upper levels. He was posted in a South American nation that wanted military aircraft. "The minister selects an 'expert' and pays him, let's say $200,000, to report on the best choice. The expert gives half of that money back to the minister," he said. "Then the companies are afraid that they might lose the contract, so they suggest a new 'expert' to look over the choices. The minister has his list. He picks one of his friends, the company pays him at the same rate as before, and again the minister receives one half. This goes on for a while." For propriety's sake, the minister does not request, and the companies do not officially pay, bribes. But everyone understands the game.

Corruption is also small-scale, and it can be measured through the experience of average citizens who are stopped for traffic violations, cited for building violations, or pressed for higher tax payments. Citizens have a *perception* of corruption, sometimes wildly at odds with statistical summaries. In Mexico, for example, citizens do not think that corruption is a problem, even though outsiders and surveys state that it is enormous. PEMEX, the national oil company, "loses at least $1 billion a year to corruption," according to the *New York Times.*[50] Drug cartels operate with police help. Millions of people who should, don't pay any income tax. The *mordita,* or bribe, is collected everywhere, and *coyotes* spirit the jobless across the U.S. border. But most Mexican citizens feel relatively safe, so they rate their nation's corruption as "low."

So do the French. But when I lived in Avignon, the left-leaning *campagnardes* of the Rhone valley routinely stole electricity from EDF, the national electric company. A woman friend showed me how she had disconnected her electric meter. A friend of hers had even made a stamp for replicating the lead seal on his meter. They think of the EDF as an enemy, because its employees have jobs for life, vacation camps, and other perks unavailable to them. In New Delhi, India, only 2.6 million of the 10 million population are registered electric users. The New Delhi State Electricity Board estimates that 1.5 million meters are altered or broken by consumers, and that it loses $250 million a year in theft. "Shanties and the homes of hundreds of thousands of people are simply hooked on to overhead electricity wires . . . as state utility workers turn a blind eye," reports the AP.[51]

Small-scale "corruption" exists in multiple cultural contexts. Hofstede and others argue that nepotism, for example, is a way of establishing trust for societies in which collective identification is stronger than individualism, in which "particularism" reigns over "universalism," and where high power distance is more common. From East Asia to Saudi Arabia and Russia, ingroups (usually based on the extended family) define themselves against outgroups, with whom they build alliances through marriages, favors, or gradual familiarity. In Indonesia, rather than a "return on investment," the ingroup looks for a "return on favors."[52] In such cultures children learn early that they must submerge personal interests to those of the extended family. What could be more natural when grandparents, aunts and uncles, numerous brothers and sisters, servants, and cousins may live under the same roof, in a few rooms? An individual who does not help his family is a bad person, so procuring jobs for relatives, directing trade to relatives, and collecting what is due in alliances outside the group is rational behavior. In many cultures, "companies" are recent (and foreign) inventions with no track records: one does business with *individuals* on the basis of personal relationships.

In Latin America this can take the form of *personalismo,* in which the peasant trades his fealty to the *lider* for the latter's patronage. This is one reason why Mexicans do not perceive the same corruption that outsiders do. The difficulty of unlearning "corruption" is often made more difficult by its invisibility. People perceive the West as requiring them *not* to look after their families, which would be shameful within the ingroup as well as outside it. What I am suggesting is that "corruption" is persistent not because it is immoral but because it as natural to most cultures as the foods they consume. The cultural embeddedness of these practices is one of the highest bars to globalization.

Smuggling and Counterfeiting

When I traveled in the Baltic states in the early 1990s, everyone was smuggling. For most people it was just a trade. Thousands of Latvians made it their business, and every morning they lined up outside the Riga central market to sell their wares. In one scheme, housewives collected "investments" from friends and neighbors, then took the train south, sometimes as far as Turkey if they were buying clothes. On return, customs officers and border guards could be bought with packs of cigarettes or the equivalent of two dollars. In Riga the women sold the smuggled goods to friends or at markets at a 100 percent markup. There was enough demand for clothing, cigarettes, and electronic goods to make smuggling an attractive occupation. By Western standards these people showed initiative, and the only one harmed was the "state," which had never done much for them and had recently disappeared. In Lithuania and especially Russia, similar systems prevailed. In St. Petersburg, the markets for smuggled goods were policed by the Mafia, which collected its own tax (reputed to be 10 percent) and determined what could or could not be sold (nothing that it trafficked in, such as gasoline, tobacco, or porn).

In Denpasar in 2001, a Balinese entrepreneur showed me his latest designs in woodcarving only after I had made several purchases and established a personal relationship. Then he led me into a locked room at the back of his store where he kept his sketches and new models. "I have three carving families in far-away towns," he said. "They only work for me. I pay them more. They must be secret on the work." If rivals saw his new designs, he said they would make copies instantly. He estimated the uniqueness of his new designs lasting two months: "First month, very good selling. Second month, okay. After that they catch up, I got nothing special."

In most of the world, there is nothing special about intellectual property. Counterfeit watches, handbags, T-shirts, CDs, and videotapes abound. If a sweatshirt sells in Poland because it has a Disney character on it, why not put that character on a juice box as well. Pirated cassette tapes of Western music were a huge business in Poland after independence and diminished only when it signed up for EU admission. I bought copied cassettes of music by The Beatles, Stan Getz, and Michael Jackson for one dollar each in Warsaw and Krakow in the 1990s.

The most lucrative piracy today involves CDs, DVDs, and computer software. The Business Software Alliance (BSA) figures that $12 billion in sales were lost in 2000 alone.[53] The BSA estimates that 94 percent of the business software used in China is pirated, a figure topped only by the 97 percent estimate for Vietnam. In Ukraine, Russia, and Indonesia, more than 89 percent of the installed software is counterfeit. Some pirates actually set up websites to *resell* the software they steal, as if they were licensed vendors.

The IFPI, a London group that represents fifteen hundred record companies, estimates that in Russia alone counterfeit CDs cost U.S. firms $1.7 billion in 2004. Russian legislator Vladimir Ovsyannikov made no bones about protecting arrested counterfeiters in an interview with the *Wall Street Journal:* "A politician isn't a politician unless he can offer *krysha* [protection] to bandits, prostitutes, or state officials, it doesn't matter. They are all his voters, after all."[54] According to the IFPI, Russia's export of pirated CDs and DVDs has increased by 30 percent since 2002, making it now the world leader. It is also the second largest market for such goods after China. And even if Mr. Ovsyannikov's pal and his goods were snatched, experience shows that 70 percent of the counterfeits end up back on the street.

No one doubts that big money is at stake. All over Asia it is possible to buy pirated copies of first-run films that are still playing in American theaters. Counterfeit CDs appear in the sidewalk markets of Beijing and Bangkok within days of their North American or European release. Counterfeit DVDs have caused Hollywood to adopt unified worldwide release dates for new movies and shortened the time that films show in theaters. In Bangkok you can find a counterfeit Gucci handbag, Minolta camera lens, or pirated copy of Word without trying very hard. In 2002 I bought Windows XP for five yuan (about sixty cents) in Guangzhou, China.

China's piracy centers on Hong Kong, which has ten times the CD-producing capacity needed for its domestic market. So where are the buyers for the 2 billion CDs produced in its eighty-four licensed factories? "It doesn't take rocket science to figure out," says Jay Berman, a London-based industry official. "This is the source of the piracy problem on a global basis."[55] Nearby Malaysia has CD-production capacity for 280 million discs, when its market is only 20 million. Taiwan and Ukraine are other major counterfeit CD centers. According to the *New York Times,* one-fifth of the GNP of Paraguay is contraband. It produces 45 billion cigarettes a year, enough for every man, woman, and child in Paraguay to smoke a pack a day. "Nearly 95% of that," writes Tony Smith, "including counterfeit versions of American brands like Marlboro and Camel . . . are smuggled through Paraguay's porous border with Argentina, Bolivia, and Brazil and then on to destinations as distant as the Caribbean, the United States and Mexico."[56]

Many underdeveloped nations lack traditions of "brands," especially foreign ones. Regional differences in rice or beer, or a particular national genius (German autos, Czech beer) are widely accepted, but the idea that brands are property is new. In most cultures with collective traditions, marques are only regional and common property. Thus anyone on Bali feels entitled to use its designs for carving, music, or weaving. Similarly, culture is common property in China. There are no enforceable copyrights or patents. An innovation that proves useful is dispersed, just as an improvement in rice farming would be. As scholars Martha Woodmansee and Peter Jaszi have shown, "intellectual property rights" are a relatively new invention.

The Western attitude toward counterfeiting was not always so moral. Not only did American film distributors pirate French films in the early 1900s, but Timothy Ryback points out in *Rock around the Bloc* (1990) that in the 1960s and 1970s Westerners smuggled jazz, rock, blues, and protest music into the Soviet Union. Vinyl records were easily confiscated at customs, so music was sometimes recorded on X-rays. It was also common for Western tourists to sell Levi's behind the Iron Curtain, and to hand out Marlboros as bribes. Ryback recounts this as a joyous process, the cultural subversion of totalitarianism, but the pigeons have come home to roost. Unless the United States and other developed nations can figure out how to protect their intellectual property in Asia, India, and Africa, they will be doing less entertaining in the future.

But it is not only entertainment that is plagued by counterfeiting. The pharmaceutical, consumer products, fashion, photography, imaging, and specialty chemical industries also face challenges. If North Korea and Pakistan can produce nuclear bombs, drugs and digital cameras will be a snap.

Taxes

When Spain imposed its first uniform national income tax in 1983, most citizens had *never* filled out any tax forms. Even in the banking and industrial Basque country, where I lived, my friends and the businessmen were flummoxed. They had always estimated what they owed and then paid up. If the taxman objected, they negotiated. They considered their Christmas bonuses, typically equal to one month's pay, entirely off the books. Only the wealthy ever had to file forms or were threatened for underpayment. In 1983 my friend the software salesman, who formerly filed what he "thought" he owed, had to take out a loan to pay his taxes. The corner coffee shop owner confided to me that he had never documented his gross sales, expenses, depreciation, or profit in twenty-five years of business. He had no idea what his building or business was worth. When the government made him put his affairs in order, he ended up paying twice as much tax as the year before. Spain was not an underdeveloped nation in 1983. In the twenty years since it began collecting taxes methodically, it has had one of the fastest growth rates in Europe. Tax compliance is one of the hallmarks of modernity.

But most local culture resists taxes. France has a vast demimonde of tax-free activity. Restaurant and bar workers, agricultural workers, truck drivers, and secretaries are often paid *en liquide*. As long as someone is working legally, a family's benefits will be the same. To *travailler en noir* has a long pedigree, as does the economy of flea markets, antique and art markets, and fruit and vegetable markets. My Avignon landlord, who owned several properties and collected all rents in cash, was a champion tax avoider. Many rural residents of the Vaucluse also avoided taxes because they raised a few olive trees or grape vines or goats, so they claimed agricultural subventions. The French government, like most, solves the problem of local tax avoidance by imposing a 19.6 percent national value added tax (VAT), by taxing bank accounts, telephones, electricity, and—most hated—a tax on the square footage of living space.

Even regimented Japan is a warren of local tax dodging. Part-timers in bars and restaurants, especially in the *mizu shobai,* instructors in music, *juku, kimono, aikido,* language and cooking teachers (especially if foreigners) are usually paid in crisp 10,000-yen notes. My friend the *mama-san* in Toyama paid her Chinese bargirls in cash every night. Tens of thousands of mom-and-pop stores maintain minimal records (older merchants in Osaka still do their accounts on an abacus!). According to the *Wall Street Journal*, "Young people who can get away with it are increasingly failing to pay into the national pension system. Among self-employed Japanese, who must make their own pension payments, 46% aged 20 to 24 didn't pay up in 2001."[57] As in France, the government compensates by having a VAT and the world's highest taxes on cars, parking, and air transport. In both nations, governments take out taxes before citizens see their paychecks.

Nations that collect taxes uniformly, whatever their systems, tend to be wealthy and to rank high on the indexes of transparency, fairness, and social welfare. At the top are Denmark and Sweden, both of which have 25 percent VATs. At the bottom are Bangladesh, Nigeria, Indonesia, and Haiti. But instituting tax compliance is far from simple. Two examples illustrate this. In 2002 Lebanon tried to put in place a 10 percent VAT to bail out its faltering economy. Anticipating resistance, it limited the tax to firms with sales over 500 million Lebanese pounds, exempting mom-and-pop stores and the merchants of the bazaar. It also excluded rice, flour, eggs, and olive oil. But merchants assumed that the 10 percent VAT would increase the rate of inflation and adjusted their prices beforehand, even the mom-and-pop stores and *bazaaris* exempt from charging it. Consumers tightened up their spending on "luxury" items, and shifted purchases from large businesses (which more faithfully pay their corporate and their employees' income taxes) to small merchants (who avoid them). So the VAT collected only 54 percent of the anticipated revenues, inflation increased, and the GNP actually declined.

The other example is Argentina. The dirty secret at the heart of that nation's 2002 default is that Argentines are champion tax cheats. "Before the value of the Argentine peso was fixed to the dollar in 1992," wrote the *New York Times*, "rapid inflation meant that delaying a tax payment would sharply lower the real value of the debt. . . . Tax-payers were not afraid because they knew they had a way out." Tax evasion made sense, and a way of life began. According to Roger Sher, analyst at the credit rating agency Fitch IBCA, 40 percent of Argentine taxpayers were cheating on taxes or avoiding them altogether by the 1990s. Sensing an approaching collapse, they shipped billions of pesos to foreign havens. Argentina already had a VAT of 21 percent, which was collected by businesses and paid to the government. They should have been collecting about $35 billion in VAT—but the government only received $20 billion. This is not only a tax evasion rate of 40 percent, Sher points out, but means that corporations collecting the 21 percent VAT were pocketing some of it themselves. After default it appeared that Argentina might institute some discipline. It received a $96 million loan from the Inter-American Development Bank just to improve its tax collections, but it still lowered its tax evasion rate only to 25 percent. Instead of simplifying its taxes, it began a system of offsetting tax credits geared to special interests, which one IMF official said "the most sophisticated tax administration in the world couldn't administer."[58]

The Resistance of the Local

When thinking about globalization, we need to take more clues from our own behavior. The United States is the most globalized nation in the world, yet we hold our local culture so deeply that we ask globalization to give us more of the same. When it comes to food, our centuries-long dependence on corn, wheat, cattle, alcohol, and sugar has changed little, despite cheaper and healthier imports and ever-mounting scientific information about our diet's danger. The processes of globalization act to give us more of what we crave: sugar and fat, now in the form of Swiss and Belgian chocolates, Australian beef and New Zealand lamb, Scotch whisky and Irish beer, French bread and pastries. But real French baguettes last only one day before becoming rock-hard, so we demand them with emulsifiers and preservatives. We can now buy inexpensive brown rice from Thailand and Japanese seaweed, which would be better for us, but we don't. Nor do we like those inexpensive, sixty-mile-to-the-gallon motorbikes that the rest of the world rides.

In language, land use, transportation, and hobbies, our practices are deeply and illogically imbedded in historic habit. Except where immigrant populations are present, we are monolingual. Second-generation Asians, Russians, Indians, and Hispanics have at best a partial command of their parents' languages. Rather than learning other people's languages, Americans work on better translation devices. Oil costs more than sixty dollars a barrel, but we continue to build freeways and Hummers, following the dream of independent houses surrounded by greenswards. In our subdivisions the houses are as big as Victorian mansions. Gun ownership, hunting, and reenactment of historic battles are popular hobbies. Bass fishing and NASCAR watching are both more popular than soccer. All around us, the local and historic persist.

Elsewhere in the world, people act similarly. Despite their unprecedented wealth, the Japanese still build new blocks of small apartments with tiny streets and no sidewalks. They still save 25 percent of their disposable income, while Americans save less than 1 percent. Despite two outbreaks of SARS, the Chinese still consume civet, turtles and toads, cats and dogs. Despite twenty-four-hour traffic jams, France still departs for and returns from the August *vacances* together. Despite Procter & Gamble's best effort, 80 percent of Europeans use aerosol deodorants rather than roll-ons. Despite dire economic necessity, in the Arab world women remain at home rather than working, often hidden in *hajib*. Female service workers from the Philippines and Sri Lanka are imported instead.

If the "American" cultural component of globalization is less powerful than critics admit, my thesis in chapter 1, it is in part because the resistance of the local is such a fragmenting force. The understanding that they have to adapt to local consumers has only recently reached the consciousness of multinationals. Procter & Gamble's manager for China said in 2004 that "you have to understand the different consumer groups, and you have to design products that delight each of them. If you take Tide, for example, the wash conditions are different around China. The quality of water that people have, the amount of detergent they use, the level of water hardness, the types of soiling are all different."[59] But Tide, he notes, competes against "many hundreds of local brands" already specially adapted. Whirlpool now designs different washing machines for Brazil, India, and China: Brazilian models sit on four legs for mopping underneath, Indian machines have a sari cycle, and Chinese customers want grease cycles, folding lids, and prefer green or gray to white enamel. It has taken twenty-five years for American companies to attain this insight.

But isn't this what *we* demand in most products? There has not been any reduction in food choice, restaurant selection, clothing variety, or transportation options in the United States. In every culture, options are expanding. Even in the most underdeveloped nations, as dire as life may be, it is no longer strictly limited to the bush or fated to be brutish and short. Globalization exists, and it is changing these worlds. But the way it does change life also flies beneath the radar of critics. It is not through the symbolic realm that globalization has its effect, but through practices entailed by new technologies in the background of daily life, my focus in chapter 3.

3 "More Than We Know"

Behind the central market in Avignon, there is an ATM on the corner of a Crédit Agricole bank. During 2002 I went there once a month, withdrawing nine hundred Euros for my rent, food, and travel. I was withdrawing funds from my U.S. bank. If I had gone inside the Crédit Agricole to exchange dollars, to cash traveler's checks, or to arrange a wire transfer, I would have paid a commission of 2 to 7 percent. Instead I paid the ATM $1.50 per transaction, an effective commission of about 0.3 percent. I could even time my withdrawals to favorable fluctuations in the exchange rate.

Later that year I made ATM withdrawals in Japan, Thailand, Indonesia, and Cambodia. Standing in a line in Phnom Penh, at an ATM made by America's Diebold Corporation, I realized that I was doing something "American," and so were the Cambodians around me. Money machines are the most visible face of what I term "easy money," an American cash-on-demand habit that flies beneath the radar of globalization critics. More than McDonald's or *Baywatch*, ATMs are changing the way people everywhere conduct their daily business. ATMs have taken us halfway to a cashless society, but here we will likely pause for some time.

ATMs

When I returned to the United States in 2002 after a two-year absence, I learned that Americans pay for 63 percent of *all* purchases with cash or debit cards. The shift is potentially a watershed, as the *Wall Street Journal* noted:

The current boom in plastic is one of those rare moments in history when agreement shifts and one payment form overtakes another as the preferred way to pay. The first such change came sometime between the 10th and 6th centuries BC when Greece and India each introduced metal coins, which surpassed barter or the shell currencies of earlier times. Coins dominated trade for the next 2,000 years, until the introduction of checks by Italian merchants in the Middle Ages. In 1690 Massachusetts became the first of the colonies to introduce paper money. Cash took decades to gain broad acceptance, but eventually became the standard of payment for the next three centuries.[1]

Now we seem to be agreeing that payment is something you swipe for. And Americans led the way to swiping.

This revolution sprang directly from the annoyance of Don Wetzel. He is the American engineer who thought up the ATM while waiting in a Dallas bank line in 1968. Wetzel was then vice president of product planning at Docutel, a manufacturer of automated baggage-handling equipment. Working with him were Tom Barnes, a mechanical engineer, and George Chastain, an electrical engineer. All their names appear on the 1973 patent. They built a prototype ATM in 1969 at a cost of $5 million, and the first one was installed in the Rockville Center branch of New York's Chemical Bank in 1973.

"No, it wasn't in a lobby," says Wetzel. "It was actually in the wall of the bank, out on the street. They put a canopy over it to protect it from the rain and the weather of all sorts. Unfortunately they put the canopy too high and the rain came under it."[2]

Wetzel and his colleagues solved this and many other problems of early ATMs. The first machines, for example, were not really connected to the bank: they were "off line," meaning that funds were not withdrawn from an account. So access was restricted to credit card holders, who used their cards as ATM cards. Back then credit cards had only raised numbers for imprinting slips, necessitating an awkward mechanical operation. Wetzel and friends developed cards with magnetic stripes and ID numbers—the present form of billions of credit, drivers' licenses, ID, and other transaction cards all over the world. The ubiquitous magstripe card was basically a spin-off of the ATM. It allowed a real-time link to bank computers, and once ATMs were connected to networks, they became popular. But making these systems work correctly, over vast distances, was still reasonably difficult. They depended on stable electric supplies and excellent phone-data lines, and they had to be restocked every day. ATMs were physical machines that were exposed to snow, rain, and sandstorms, to blistering heat and subzero temperatures and thieves. ATMs had to do it right or customers would lose confidence. Making them work overseas, via deep-sea cables and satellite, depended on a logistical wherewithal that was peculiar to America.

The goal was to connect the machines simultaneously to the ATM network, the credit card (POS) network, the Automated Clearing House (ACH) network, and the bank's cash network. This complete integration was achieved finally in 2003. In theory, as Wetzel noted, "you could buy stamps and concert tickets, you can even trade stocks" on them.[3] The technology developed for ATMs underlies debit cards and other forms of "easy money." But because America's use habits followed the ATMs abroad, only in Japan and more recently India do people use these extra capabilities.

Not until 1988 were there 100,000 ATMs installed, but the next 100,000 came in only four years. In the same year that Wetzel invented the ATM, Japanese firms figured out how to bond a magnetic strip into a bank passbook, developing a rival system. This machine-readable passbook system is still common in Japan. As personal computers increased in power and declined in cost in the 1980s, and as fast, cheap, reliable modems became available, the cost of an ATM fell to under $20,000. By the mid-1990s, even lower computer prices and advances in CPU speed cut the price in half for ATMs installed inside grocery stores, shopping centers, and convenience marts. At $7,000 to $10,000 per unit, these ATMs cost only 20 to 30 percent as much per year as one bank clerk, and they never went home.

More than 1 million ATMs had been sold worldwide by 2002. U.S. banks alone spent more than $1 billion on them. The Asian-Pacific market by 2002 had more machines per square mile than Europe (second place) or North America (third). Central and South America lagged behind, and ATMs had penetrated relatively little of Africa (in some countries there were still none in 2004). By 2003 Visa-branded cards were accepted at more than 480,000 ATMs via the Visa Global ATM Network, and MasterCard's Cirrus network had more than 350,000 machines in foreign countries. Both networks have machines in more than a hundred countries. Using interbank alliances, Visa and MasterCard holders can use more than 1 million ATMs worldwide.

Docutel is long gone, but American firms still lead the world in ATMs. Since 1990 Diebold has produced ATMs in a joint venture with IBM called Interbold. In 1996 Diebold began manufacturing and shipping thousands of machines itself, marketing a "Cashsource Plus" machine outside its IBM agreement. NCR (formerly National Cash Register) is another significant presence, but its sales figures are not published. The third-place firm is Triton, also American, with about 12 percent of the U.S. market. Triton is the fifth largest ATM manufacturer in the world, specializing in "off premises" machines. Other American manufacturers include Greenlink and Tridel, which builds the most common grocery store ATMs.

As the technology matured, foreign competition developed. Japan's Fujitsu ICL commands 34 percent of that archipelago's ATM market and a comparable portion of the Southeast Asian market. Hitachi is a rival, as is Multimedia Business Machines (MBM) of Malaysia, which undercuts Japanese firms and provides multimedia, local language interfaces in, for example, Bahasa Malaysia and Tagalog. In Europe, the venerable Bull PLC of Britain is the leader at home and in France, Austria, Hungary, and (it claims) third in China and Taiwan. The Banqit firm leads in Sweden, while Wincor/Nixdorf (formerly part of Siemens) dominates Germany and Switzerland and has machines in the United States. Daussault and Olivetti also manufacture ATMs in Europe. In short, the United States no longer dominates the technology it invented.

But U.S.-style ATMs have withstood major technological challenges. The European "smart card" system, in which a microchip is implanted in a cash card that contains vital user data for identification, was widely used until 2001. During that year the system was breached in France. Hackers reverse-engineered a few smart cards, bought thousands of new blank cards from manufacturers, and sold their clones in the same Paris *banlieu* where stolen cell phones were for sale. A programming change ended this problem, but now users need a code, like Americans, and they still have lots of personal financial data on perishable, easily stolen cards.

Interfaces now vary somewhat around the world. European ATMs have interfaces in several languages, typically English, French, German, and either Spanish or Italian. They accept multiple European bank and credit cards, as well as Visa and MasterCard. Since 2000, most European ATMs have also accepted U.S. bank cash cards if they are networked through the Visa Global or Cirrus systems. Although this is not particularly advantageous to European banks, which collect only half the access fee (and no exchange commission as of 2002), it is very advantageous to European businesses and thus to the banks indirectly. If tourists can easily spend their money in local stores, the spending of local merchants (who are bank clients) revs up the velocity of local money.

By contrast ATMs in the United States are not so foreigner friendly. They may have a Spanish-language interface, but they seldom reciprocate with European ATM networks. Only in 2004 did Citibank begin reciprocal ATM relations with affiliated Mexican banks. As of 2004 most bank cards from Japan, the world's number two economy, do not work in U.S. ATMs (Japanese banks chose to make them incompatible). On the other hand U.S. ATMs are theft-proof and built to withstand Minnesota winters or Arizona summers. U.S. firms pioneered drive-through, convenience mart, and grocery store ATMs, raising the expectation of "cash-as-you-go" to a level unknown anywhere else until recently.

In much of Asia, the local language is usually the only language on the ATM, and it may read top to bottom or right to left. Bilingual ATMs (usually local language and English) are common in tourist destinations and large cities. Rather than buttons, touch screens are the norm in Asia, as are passwords. ATMs are seldom outside, where they would be exposed to monsoons. In Japan most ATMs shut down around midnight, so that no one can withdraw a large sum of money for an irresponsible spree. However, Japan's users can pay bills and buy tickets or even mutual funds at their ATMs. The Japanese are the world's leading tourists in number and spending, so why don't most Japanese bank cards work in foreign ATMs? Well, there is cultural resistance to "easy money" in Japan. "Plan your entertainment carefully and don't spend more than you should" is Japan's advice to carousers and tourists alike.

In 2004 most of the world's ATMs still dispensed only paper money. Ninety-five percent of the 220,000 ATMs in the United States spit the folding green alone, and of the 8.3 billion transactions that we performed at ATMs, more than 90 percent were cash withdrawals. Using an ATM saves a great deal of time, as those of us who stood in line on Friday at banks will remember. It saves banks money, increasing productivity. Above all, it gives us our money *easily,* when we want it. But the world is gradually customizing ATMs for local use.

The instrumentality of the cash machine is that it eliminates the need for advance planning for discretionary spending, such as movies, markets, small purchases, and perhaps dinner. In 1980, if I knew my weekend included a number of cash-only activities, I withdrew the funds on Friday. I could not spend more than my withdrawal. By 1995 I had no need to plan. The ATM made spending always possible. It made one more bar, last-minute theater tickets, and craft market purchases ordinary. There were (and still are) many people who did not have credit cards and did not carry cash; ATMs give them flexibility and safety in shopping. To look at the other side of the phenomenon, the ATM has promoted many cash-only businesses and is hastening the demise of the check.

The effects of the easy-money revolution are more dramatic overseas. There ATMs do not piggyback on either a history of consumer credit or on the widespread use of credit cards, because fewer merchants have ever accepted credit cards. Today, however, ATMs help to sustain small merchants in the pedestrian zones of Avignon, Kyoto, and Florence. They fuel the lunch trade in Rio, and resorts from Dahab, Egypt, to Bali, Indonesia. In these areas, proximity to an ATM has supplanted proximity to the bank as a desideratum of commercial location. Many big banks in these nations often have no street-side ATMs, to the detriment of local businesses. But I have seen the tipping point in Indonesia and Egypt, when locals began to use ATMs themselves. The ATM-ed neighborhoods grow, while the rest of Egypt or Indonesia lags behind. Even a single ATM can power a shopping or a nightlife district. These are echoes of a global trend, which is to locate ATMs in airports, grocery stores, shopping centers, in Virgin Records or Starbucks, creating consumption hubs apart from physical banks. Futurists may see this as a halfway measure to universal debit cards, but technology is likely to pause here, because the teller machine is now debugged and basically foolproof.

Several nations have tried to ATM overnight, with mixed results. South Korea had the stable electric system, telephone lines, and computers in place when it made the leap in 2000. But it did not have experienced banks, sound accounting practices, or cultural experience with consumer credit. To break its dependence on exports, South Korea wanted to rev up its domestic economy with easy money. It encouraged ATMs everywhere, required most retailers to accept credit cards, and gave tax deductions to consumers for purchases made with them. The domestic economy boomed, but neither banks nor consumers knew how to rein in personal credit. Bankruptcies and bad loans leaped in 2002 and 2003, causing a nationwide retrenchment. Nonetheless, by 2004, 87 percent of Seoul stores accepted plastic. Swiping had become normal.

China decided to follow the same path, wanting a credit card system in place for the 2008 Olympics and 2010 World Exposition. In early 2004 Beijing officials told local merchants that 50 percent of them must accept credit and debit cards by the end of 2005, and that 90 percent must comply by 2008. China wants to capture the spending of visitors. But it faces greater obstacles than South Korea. It lacks telephone infrastructure, and only 1 million Chinese had credit cards in 2001 (0.25 percent of the population), though another 24 million carry debit cards. Only 4 percent of the nation's merchants accepted either. Such cards accounted for only 2.7 percent of overall consumer spending in 2001, and only 5 percent of cardholders carried a balance (in the United States 75 percent of cardholders carry a balance at least once a year). Furthermore, credit cards issued in commercial south China don't work in diplomatically oriented north China. In fact, many bank cards don't work at all outside the town of the issuing bank. There is no card reading standard, and instead of importing a foreign one, China will try to develop its own. Outside of major cities, ATMs are scarce, and Chinese banks are already staggering under bad debt. But China will be ATM-ed, because it sees the correlation between prosperity and the ATM-credit society. A Nielsen Report on the Visa and MasterCard credit and debit card charges for 2002 makes this clear:

1. United States $1,200 trillion
2. United Kingdom $247 trillion
3. France $164 trillion
4. South Korea $104 trillion
5. Canada $90 trillion
21. China $13.6 billion

Given the size and growth rate of the Chinese economy, the level of its domestic spending is low, and there is no doubt that ATM, credit, and debit cards boost domestic spending. Already younger Chinese are gravitating to "easy money" culture. Ye Rong, a Shanghai newscaster quoted in the *Wall Street Journal*, says she uses her credit cards daily, "not only for big things, but for clothes and even food."[4] College student Mao Jia Yi plans to get a card as soon as she qualifies: "Then when I want to buy something, I can just buy it." But to become widespread in China, ATMs will probably have to be adapted further to local customs.

For a glimpse of adaptation, China can look to India. There the ICICI Bank and HDFC Bank have pioneered efforts to use ATMs for small accounts. Neither bank existed in 1990, but by 2004 both had millions of customers and profits that had grown by 30 percent per year for five years. They achieved this growth by reconfiguring ATMs for people who made less than $100 a month, many of them illiterate. Larger banks thought that these customers weren't worth the effort, because they usually had only the equivalent of $20 to $30 to start an account, maintained a low balance, and dealt in tattered ten-rupee notes (equal to 22¢). Big banks calculated they would lose money on such customers.

These poor Indians had to stand in long lines to cash a check or to pay a phone bill before 2000. Clearing a check or transferring money required a week or more. There were some ATMs, but the phone system was so bad that three lines had to be attached as backups against disconnection. Nor was telephone banking a possibility, because most of India has a rotary phone system rather than touch-tone. Guards had to be posted at ATMs, and they spent a lot of time showing people how to swipe their cards. Then ICICI Bank, with the help of the Indian Institute of Technology, developed an ATM that could be built from local parts for $800. It's not as weatherproof as Wetzel's, but it works. A special bill scanner and software solved the ragged rupee problem. Instead of opening bank branches, HDFC Bank placed ATMs in stores and made accounts available through Internet kiosks. ICICI even has ATM vans that tour remote neighborhoods of Bombay. Fingerprint scanning allows the illiterate to log in, and customers are encouraged to use cell phones to bypass the antique rotary system. In fact, a new account based on cell-phone-only access can be opened with a $20 deposit. These accounts have been wildly successful, growing by 60 percent in 2003, and now number more than a million. "Before you would not find 10-rupee notes in any ATM anywhere," says Pavrita Ramanujam, a project leader in the ATM effort, "even though rural people needed them. We have built the system from scratch because there is no easy access to banking in remote areas."[5]

Among other India-specific adaptations: account holders can make donations to their favorite Hindu temples, and even get a receipt to present to their gods when they go in to pray. They can pay their electric and telephone bills, buy and sell mutual funds, or top off cell phone accounts.

A different version of easy money has appeared in Hong Kong. The MTR Corporation that runs the city's subways simplified the crush at its turnstiles with stored-value cards similar to those used on some U.S. tollways. Stored value cards are not conceptually strictly the same as ATM cards, but what I am examining here is *use habits*. In fact, these cards have become so popular that they have migrated upward to become cash or credit cards. They now compose 2 percent of all cash transactions made in Hong Kong and their use is spreading elsewhere rapidly. The "Octopus" card contains a chip that can be read through leather and plastic by a low-range radio transmitter, so users can just wave their wallets over subway turnstiles. The cards can also be used at parking meters, at Starbucks to buy coffee, at 7-Eleven to buy newspapers, at municipal swimming pools, and even at racetracks. The Octopus can be recharged at the MTR office or the bank, and of course at any ATM. Twelve thousand locations across Hong Kong accepted the card by 2004. The similar Singapore CashCard, developed in 1996, was only slightly behind with ten thousand participating locations.

The ATM phase of easy money now has great "momentum." This is a phrase that historians of technology use to describe devices that business and government have massive investments in and commitments to, devices that may not be perfect technically but have habituated everyone to use them efficiently. By 2003 ATM's software had such a tight fit with the ATM machinery that it was effectively optimized at the performance checkpoints of speed of response to the customer, security, and accuracy. And ATMs had taught users to employ them according to their strengths. For example, they don't accept or dispense coins, which is hastening the decline of metal money. They have made U.S. twenty-dollar bills an international currency, so important that these had to be redesigned to prevent counterfeiting internationally. ATMs have spawned their own slang, such as "ATM disease" (having no change, but a wallet full only of twenty-dollar bills, which are in turn known as "yuppie food stamps"), to ATM someone (hitting or groping below the belt), and "drink link" (an ATM machine).

Over the medium term, ATMs will spread American attitudes toward the ready availability of money. Over the long term, they will prepare us for the step to "credit culture," about which much has been written. But, as the earlier quote from the *Wall Street Journal* suggests, shifts in money occur slowly. A "global credit culture" may be somewhat distant, but "easy money" culture is here: it is the American attitude that you should have access to all your money all the time.

The Money Market

One hundred thirty years ago Walter Bagehot, writing on the origins of the money market, pointed out that British colonialism and trading culture had created something he believed to be culturally specific:

> If you take a country town in France, even now, you will not find any such system of banking as ours. Cheque-books are unknown, and money kept on running account by bankers is rare. People store their money in a *caisse* at their houses. Steady savings, which are waiting for investment, and which are sure not to be soon wanted, may be lodged with bankers; but the common floating cash of the community is kept

by the community themselves at home. They prefer to keep it so, and it would not answer a banker's purpose to make expensive arrangements for keeping it otherwise. If a 'branch,' such as the National Provincial Bank opens in an English country town, were opened in a corresponding French one, it would not pay its expenses. You could not get any sufficient number of Frenchmen to agree to put their money there.[6]

The growth of the British East India and Hudson Bay companies in the 1600s and 1700s created the need for a way to settle accounts with distant trading partners. "A large amount of money is held there by bankers and by bill-brokers at interest. This they must employ, or they will be ruined," Bagehot noted. This is among the first observations of the pressure that bankers feel to wring the maximum from every dollar on deposit. Americans may take this for granted, but it is relatively new in much of the world, and it's a kind of "technology" that is reorienting many people's habits.

While the mechanics of deposit banking long ago spread to non-Anglophone worlds, only at the end of the Cold War did the pressure to maximize return on deposits become worldwide. Through the 1970s, most nations maintained some kind of de facto bar to unhindered trade in deposited funds, and a lot of people still hid their money under the mattress. By the 1980s these barriers were falling, and relatively inexpensive computers connected to intercontinental cables made it possible for institutional investors and then individuals to move money over oceans to take advantage of the best returns. The United States became the world center of deposit banking, currency trading, and other "financial instruments." Those who worked this way with much "fast money" were memorably labeled the "Electronic Herd" by Thomas L. Friedman:

> There suddenly emerged a vast global plain where investor herds from many different countries could roam freely. It was on this wide-open plain, later expanded into cyberspace, that the Electronic Herd could really graze, grow, multiply and eventually gather in powerful Supermarkets. By the late twentieth century the dominant fact of the global financial system was that the private sector had become . . . "the overwhelming source of capital for growth," replacing the public sector. This has been true both within countries and between the developed countries and developing countries. According to the U.S. Treasury, in

the 1990s nearly $1.3 trillion in private capital has flowed to the emerging market economies, compared to roughly $170 billion in the 1980s and a relative pittance in the 1970s.[7]

The "herd" metaphor fails in some respects, and in fact, there's a great danger in applying metaphors to capital, especially to personifying it. The money market is no single entity, but many movements occurring simultaneously. Some capital is prospecting for new opportunities, other capital is tirelessly planting and harvesting old fields, and some capital is stagnant, dissipated, lost, stolen, or exchanged for goods that do things. It is constantly changing form, as hedge funds outshine derivatives, and "private equity," called "the new king of capitalism" by the *Economist* in 2004, takes the luster off both. Here I use the term *money market* in the broad sense, to indicate the now international expectation that one's liquid assets should be earning interest rather than gathering dust under the bed. Ultimately these funds get invested in something, and to a greater degree than ever before that something is overseas. Economists call these overseas investments foreign portfolio investments (FPI). These are different than fixed assets called foreign direct investments (FDI).

Today everyone is in the money market, from Silicon Valley investment bankers to insurance companies to TIAA-CREF, from Crédit Agricole and General Electric to Australian cattle ranchers. Most have their money "secured" by stocks, bonds, certificates of deposit, futures, currencies, and commodities. Everything from mortgages to shopping centers, from student loans to the price of gold in 2010, from United States debt to Jim Jarmusch's next three films has been "securitized." David Bowie set a trend in 1997 when he traded a piece of his future royalties for $55 million in securities called "Bowie Bonds." There has been an expansion of this kind of investing coincident to the aging of the U.S. baby boomers, who sought wider opportunity and accepted the risks of leverage. Companies that once paid the mortgages on the buildings they occupied, now routinely turn that debt into REITs (real-estate investment trusts)—for others to invest in—and thereby free up their own cash for other purposes. The adventurous can invest in the potential appreciation of Barry Bonds's home run baseballs or *grand cru* wines.

The trading face of the money market is international banks and currency traders. They scan the globe for opportunities or disparities that allow them to leverage their huge assets in different markets simultaneously. They have computer models to spot small market misalignments. But their funds actually belong to professors from Duke and little old ladies from Dubuque. By 1992 American assets in mutual fund retirement plans alone reached $412 billion; five years later the figure had quadrupled to $1.6 trillion. The impetus for this growth was the Individual Retirement Account. By 2000 more than one-half of the U.S. population was directly or indirectly invested in the equities market. Mutual funds themselves are not making currency hedges, but they are invested in banks, securities firms, and investment firms that carry on this business. By 1997, according to Saskia Sassen, 83 percent of the world's equities under institutional management were traded on twenty-five major markets.[8] Only about 10 percent of investors were buying foreign stocks directly, even though Toyota, Nestlé, and BMW would seem obvious choices. On the other hand, Sassen notes that this capital under institutional control represents only half of global market capitalization—there is another $21 trillion that is direct investment by transnational corporations (FDI).

Information is key to the money market—and it expanded greatly as money came out from under the bed. Friedman recounts that in 1976 Goldman Sachs's securities analysts each covered six different industries and seventy-five companies. But by 2000 each analyst covered only twelve companies in one industry, as it became important to look into each company as deeply as possible. "In the old days, a stock might have stayed at $4 or $5 for a while," says Jeff Van Harte of the Transamerica Premier Equity Fund: "Now all it takes is a couple of calls to hedge funds with people saying the business looks good, and it's at $9 in a couple of trading sessions."[9] Indeed in the 1990s the S&P 500 averaged just nine days a year when it rose or fell more than 2 percent. That figure has increased to thirty-one days over the past five years. Computer models of future economic conditions, plugging in variables as diverse as weather conditions, the spread of hoof-and-mouth disease, or increased human longevity due to new drugs give traders more information. If they can predict that a series of hurricanes will lash Florida, they invest in Wal-Mart and Home Depot. If they suspect that a new feed additive will reduce swine mortality, they sell short futures in hog bellies. If they see cell phones catching on in Vietnam, they buy shares in NTT Dokomo, the dominant cell phone company in Asia. But they still miss the crises, such as Enron and Parmalat, not to mention Hurricane Katrina.

When something goes wrong, the individuals recognized as "players," mostly American-based financiers such as George Soros or the Bass Brothers, have been ridiculously demonized. Public opinion seems to need a human face and narrative to explain sudden money movements.

The Bass Brothers have lost big twice, once in a failed attempt to dictate the market in silver futures in the 1980s, and a second time in 2001 when the fall of the stock market precipitated a margin call and they had to sell a huge chunk of Walt Disney Company. But they are remembered only for "cornering the world market in silver." Foreign banks are more common among currency cowboys than American moguls. The national bank of Malaysia lost $3 billion speculating on the British pound in the early 1990s. Prime Minister Mahathir Mohammed blamed a "Jewish conspiracy" and Soros for the losses. Other traders thought the accusation smelled fishy. For them the message was clear: if the Malaysian Central Bank is speculating in the pound, who knows what other foolishness is afoot. So they pulled all their money out of the Kuala Lumpur stock market, causing it to drop 48 percent in 1998. The Malaysian ringgit fell to a twenty-six-year low. In the same year that Russia defaulted on its bonds, the world learned that South Korea, another developing nation, had invested billions in those bonds. So had Thailand. And Malaysia. It was a get-rich-quick club. Large Western banks were also caught, but most had protected themselves with default guarantees from their governments. Once a currency like the Malaysian ringgit or bonds from Russia begin to fall, there is no ready market in which to sell them. There is not enough appetite or enough trading depth in Kuala Lumpur or Moscow to absorb billions of dollars worth of credit, especially if the fall is likely to continue.

The most important cues about money markets still come from *local reception.* The U.S. money market, like other aspects of its economy, is highly scrutinized. European economies are almost as transparent. Asian economies, even that of Japan, are somewhat opaque, especially to outsiders. Thus New York currency traders are not first to know if inflation has picked up or the black market currency exchange rate has changed. Local shoppers know first. The local restaurateur can sense the amount of currency in circulation when he goes to the bank to get change—if the bank has trouble finding $300 in change, maybe there is a liquidity crisis, no? The local tax collector knows if he is collecting more or less than last year, and hence has a clue to budget shortfalls. Local money changers know if remittances are converted into the local currency, or if locals are changing every peso into Euros and dollars.

The dirty secret about currency devaluations like those in Mexico and Argentina is that local people knew long before outsiders that they were about to happen. "They begin with a local banker," notes Friedman, "moving his money out of a country by converting from his local currency into dollars or betting against (shorting) his own country's currency in the forward market." In fact, an IMF study published on the 1994–95 Mexican crash (published in 1998, when all the transactions could be reconstructed) concluded that "domestic residents and not international investors" started and sustained the rout. I was in Mexico just before the crash, and many restaurants and stores requested dollar payment. I had the same experience in Russia in 1996: no one in St. Petersburg wanted rubles, and no one honored the official exchange rate. As Thomas Friedman wrote, "the first bull is always a local."[10]

Mexico defaulted in 1995, but promised to put its house in order, and by 2000 it was back in the good graces of international investors. Why? Among other changes, it agreed to make data on the money market more immediate and more transparent. South Korea was near bankruptcy in 1997, after the Russian bonds and other chicanery, but by 2001 it figured among the Asian Tigers and looked good to money managers. It wrote down the fictive assets of many big banks and *chaebol*. Money market managers believe that they won't get burned in *that way* again, and in fact they usually do not. When Argentina collapsed at the end of 2001, most of them escaped. They had learned to monitor local investors, who had been deserting the Argentine economy for two years. "There is really no one who will have been caught off guard by events there," said a Bank of America official in London.[11] The only effect of the Argentine debacle was to depress the Euro slightly, because many of the banks holding Argentine debt were Spanish (an example of enduring cultural affinities). The size of Argentina's problem became known when Domingo Cavallo, the economics minister who had worked tirelessly for financial transparency, said he "could not tolerate the corruption in government" anymore and then quit.[12] Then everyone knew that Argentina was about to take a dive.

The result of this period of crisis has been even greater overseas investment, but in a changed form. Where a Swiss or American company formerly invested in Mexican or Argentine companies or their stock, a foreign portfolio investment (FPI), they have now changed to foreign direct investment (FDI), actually owning in-country assets. This is a way of bypassing some of the problems of lack of transparency in local corporate law, accounting, and other practices that had made overseas markets so volatile before. But FDI also makes the foreign investor much more committed to staying.

Both kinds of capital have affected the developing world. They have made capital available. "Poor countries, with large investment needs," the *Economist* wrote, "are no longer hamstrung by a lack of capital. Savers are not confined to their home market, but can . . . seek investment opportunities that offer the highest returns around the world."[13] Friedman seems not to be exaggerating when he writes that "this diversity of investment instruments and opportunities has been a godsend for both developed and developing countries and companies." We forget that no one used to invest in underdeveloped nations, but now the citizens of Dubuque do so. William W. Lewis has documented how Brazil, India, and Russia benefited from direct investment when development economics of the sort backed by international aid organizations had manifestly failed.

But the fluidity of global capital flows has also changed the United States. Firms such as General Electric have moved into the credit markets of nations such as Thailand (through its Capital OK). Private equity firms such as KKR, once regarded as pariahs, have become corporate "problem solvers." Sony employed the services of two of them to buy MGM recently. Private investors put $100 billion into these private equity companies in 2004, up from only $10 billion in 2000. It is a "godsend" for Americans in the sense that it induces foreign nations and individuals to soak up the enormous debt of the United States, keeping interest rates low and Americans in their houses.

Flexible Manufacturing

Where are Xboxes made?

When Microsoft rolled out the Xbox in November 2001, promising to spend more than $1 billion to compete with Sony and Nintendo, many people thought it would be a boon to U.S. electronics manufacturing. But then they asked, "Where will it be made?"

The first Xboxes were made in Guadalajara, Mexico, by a Singapore-based company called Flextronics. Its shares, traded on the New York Stock Exchange, rose by almost 30 percent in the following month. The Microsoft-Flextronics arrangement offers a glimpse of how U.S. companies, trying to keep their production nimble, have enriched the developing world. It's not outsourcing because Xboxes never were made in the United States, and production can leave Guadalajara faster than *mariachis* quitting a cheap *cantina*. It's flexible manufacturing.

According to its prospectus, Flextronics operates more than a hundred "flexible assembly" plants in "low cost" countries. It owns some, and it leases some in partnership with local investors and governments. In 2004 it bought the manufacturing operations of Nortel, one of Canada's largest companies. More typically though, Flextronics leases its equipment and subcontracts services such as delivery to others. It constructs plants so that they can be converted from the production of Xboxes to cell phones or laptops in a matter of weeks. It has "core employees" who can teach local employees the skills to build new items in on-site labs. At each site, it builds a network of local suppliers for plastic parts, trim, packing material, and whatever else is needed.

Flextronics did not invent this model, though it has aggressively developed it. This is the result of an evolution that began in the bicycle industry in the United States in the late 1800s, which Henry Ford took for his factories, where it became known as "Fordism." The bicycle makers were the first to move manufactured goods along an assembly line of specialized tools and sequential operations, with distributed overhead power. Some scholars still refer to this as Fordism, but the model was superseded quickly by that of General Motors, which pioneered the annual model change, while Ford was still churning out the same black Model T's every year. Fordism also involved a great deal of redundancy: annealing vats and milling machines, for example, were supplied at every point in production where a part needed tempering or grinding. This sped up output, but it was capital intensive and inflexible. Except as a metaphor, Fordism has been dead since 1927, when Alfred P. Sloan began the yearly model change at GM, and Charles Kettering pioneered such consumer-friendly features as electric starters and headlights, not to mention colored enamel paints. The GM factory floors were set up to produce cars in different colors, with different options—and the whole assembly line was reconfigurable when new models came out. GM's scheme produced Chevrolets in a rainbow of colors in 1927, putting the black Model T out of business. It was the dominant production paradigm for fifty years.

Around 1975, however, there began a shift to "lean production," which was associated with Japan's Toyota Corporation but incorporated many American efficiency advances. Then Sakichi Toyoda and his son Kiichiro established the *kanban* (literally, pull-tag) system of production, with its principles of *jidoka, kanban,* and *kaizen* ("just in time," "pull tag," and "constant quality improvement"). Lean production could not have developed in the United States in the 1960s or 1970s, because America's mind (and hence automotive style) was in spasms of excess: think of Buick Roadmasters, Jayne Mansfield, and Liberace. But lean production blended the American ideas of product flow and flexibility with the Japanese genius for parsimony, worker consultation, and social harmony. The original Japanese *kanban* model depended on extraordinary coordination with suppliers, while the United States was traversing an era of protests and rebellions, not to mention antitrust litigation. Toyota suppliers essentially could not work for other firms because they possessed trade secrets. In return for their fealty, they had to be paid above the market price. Toyota Auto, itself a subsidiary of Toyoda Loomworks, would go on to create Toyoda Electronics, which was partially owned by Matsushita, which could not then work with Honda. In the United States these firms might have been sued for collusion. But in Japan the model for this kind of trade dates to the feudal past. More antithetic still to the U.S. climate was the obsession with constant product improvement through customer and worker feedback. John Dower has written in *Embracing Defeat* about the attitudinal shifts in the Japanese population during the U.S. occupation that set up such innovations as the minimal inventory and "compressed warehouse" concepts.

We often forget that the first Toyotas to arrive in North America were miserable cars. The 1960s Toyota Crown had virtually no suspension, a body that rusted at the sight of salt, and a small, high-compression engine that quit after forty thousand miles. But the Japanese studied the American market and adapted to it. Unlike GM they could not mass-produce cars and foist them on consumers with incentives. In fact, they had to pay very high tariffs. As they perfected their myriad of products for export, they had to wait for them to catch on. These products were "pulled" through the Japanese manufacturing system largely by American consumer demand, which resulted in steadier relations between workers and employers and suppliers back in Japan.

It would have been difficult to transfer all *kanban* techniques to America, because of their cultural embeddedness. After two decades of effort, the *kanban* "masters" at Toyota's Georgetown, Kentucky, plant are still hampered by their limited English skills and their workers' lack of an intrinsic *kaizen* mentality. Most them leave their families in Japan so they won't become "too Americanized." But American manufacturers caught on, lifting those pieces of the *kanban* system compatible with U.S. business and worker culture. The idea of just-in-time delivery made sense: reduce inventory, only produce enough to replace supplies, eliminate warehouses, increase product and tool flexibility, and speed up delivery. Americans set up new inventory, accounting, and cost systems. These won converts and then became a competitive necessity. The web of producer-supplier relationships around Detroit, for example, had been geographically limited to a few hundred miles. But with airfreight, that was remapped to include the southern United States and later Mexico. Hundreds of small companies in Michigan, Illinois, Indiana, and Ohio that once produced GM moldings or Ford taillights on sweetheart contracts suddenly had to compete. They had to reengineer their products constantly or be replaced by rivals elsewhere. As flexible manufacturing grew in the United States, new companies were founded on the model. Nike, for example, never had production facilities in the United States. It made shoes all over Asia, largely through South Korean subcontractors who moved from country to country, seeking the best tax rates and lowest labor costs. Cisco Systems, which makes Internet routers, produces its goods at thirty-four foreign locations, of which it owns only two.

The stars of flexible production—Flextronics, Solectron, and Celestica—are companies basically focused on the process rather than a specific product of their own. They added to the model the completely reconfigurable factory floor. In theory no plant need ever close. They can produce any electrical item, from a toaster to a network server. These companies gamble that cutting downtime will recoup the cost of giving their employees new skill sets. Their local supply alliances will allow them, they say, to turn on a dime as demand changes. If there are downturns, they fire a few hundred employees here and hire a few hundred there, moving work around. This is preferable to closing entire plants or the company itself failing, which is the sort of catastrophe that used to happen (although these firms have closed plants). Microsoft likes this setup: it does not risk huge sums in building plants. Its risk is diluted through a subcontractor who has countless subcontractors. If the Xbox is a failure, Microsoft loses its research costs, but it does not own a defunct factory. Cutting-edge U.S. companies have embraced this model, contracting out significant portions of their manufacturing, both at home and abroad.

In fact, the popularity of the Xbox allowed Flextronics to expand production in 2001 to its plant in Hungary. Then in 2002 it shifted Xbox production to its plant in Doumen, China, to be closer to its sources of chips. Were the Hungarians thrown out of work? No, they began to assemble television sets for the Chinese company TCL, to be sold in European markets. Flexible manufacturing gives consumers in the developed world a stream of new products at low cost, but its greatest benefit is that it has raised the standard of living in much of the underdeveloped world.

Franchising

When we entered that KFC in Nishinomiya, back on the first page, we entered a franchise, a form of business hardly known outside the United

States before 1950, but one that has mushroomed in Asia since the 1970s. A *franchise* was originally a grant from a local lord to hold markets or fairs, to operate ferries, or to hunt on royal land. Then kings got into the act, granting franchises for all manner of commercial activities, from building roads to brewing ale. The first private award of a franchise occurred in the 1840s when German brewers granted certain taverns the exclusive right to sell their beer. In 1851 the Singer Sewing Machine Company began to award franchises to dealers in the United States. By 1900 U.S. oil refineries and auto companies were franchising dealerships, and they were followed by soft-drink makers.

Often franchises were the poor man's gangway into business. Harry Oltz was an unemployed paperhanger in 1937 when he invented the machinery for soft ice cream. Disdained by established creameries, he eventually franchised 3,750 Dairy Queens. William Rosenburg drove a lunch wagon before he started Dunkin' Donuts. In the United States franchises fit in with the preference for suburbs and auto travel—which may account for their growth here. They developed not in large cities, but in places such as Waltham, Massachusetts; Des Plaines, Illinois; Arlington, Texas; and Norwood, Ohio. They capitalized on convenience. There were already many successful franchises in the United States before World War II, but their growth exploded in the 1950s and 1960s.

Today there are many forms of franchising. Among them, one common and "American" feature is a very effective middle-management layer. Unseen by customers, the franchisers' middle management advises franchisees on how to manage their images, employees, and materials, and how to work within the dictates of the franchiser, who coordinates the franchise image nationally through advertising and marketing. There are meetings, newsletters, videos, training, campaigns, and the "pep" that novelist Sinclair Lewis made fun of in the 1920s. But Americans do this very well, and their ability to standardize the practices of a decentralized business operation has had an enormous impact on globalization. Franchising creates business activity, reduces risk for franchiser and franchisee, guarantees cash flow, and attracts customers through standardization. It is no exaggeration to say that franchising is one of the most powerful creations of modernity.

The first important form of franchise in the United States was the "name and process franchise," the original system of Kentucky Fried Chicken and One-Hour Martinizing. The franchisee could use the processes and recipes, the names and logos. The current model is called "business format franchising," and is much more complete:

Specifically the franchiser transfers all its operating systems, technical expertise, marketing systems, training systems, management methods and essentially all relevant information, to the new franchisee. The franchiser also trains the new franchisee extensively up front and provides ongoing training and support throughout the life of the franchise agreement. Business format franchising is what franchising is all about today and is essentially why franchising is the most successful method of distributing goods and services in the economic history of the planet Earth.[14]

That's the International Franchise Association speaking, and the claim may be truer than liberals care to admit. From auto dealerships to fast food to clothing stores, from soft-drink bottlers to real-estate agencies, franchising is so central to modernity that many people mistake it for "business," when it is only one form of commerce.

To illustrate how franchises rise on the standardized practice of decentralized business operations, let's take the famous case of Ray Kroc, a milkshake machine salesman. He merely *bought the right to franchise* the hamburger business of the McDonald brothers of San Bernardino in 1954. They had attracted his attention because they bought so many machines from him. They had a great business in low-cost burgers, but no ideas for expansion. The notion of replicating a successful business plan elsewhere was beyond them. For many years it was beyond Kroc too; he made more money at his own McDonald's in Des Plaines, Illinois, than he did in franchise fees. But he relentlessly standardized his concepts of cleanliness, food preparation, and uniformity of appearance and service. He sought to sell his system; he did not seek to own outlets initially. Slowly they caught on. With associate Harry Sonnenborn, he pioneered site selection and real-estate leasing for his franchisees. They were wary of big cities and alert to the increased personal mobility of Americans after World War II. Eventually they had thousands of franchisees, whose restaurants, staff, and food met the same, relatively high levels of speed, cleanliness, and quality. It took decades, because the idea that franchise equals uniformity was not part of consumer consciousness.

Most franchises didn't start out as big business, but as the dream of some little guy like Kroc. Getting a franchise in the early days was the way to leverage a few thousand dollars into a real business through long hours, to achieve middle- or upper-class status. Franchisees were risk takers, frequently mortgaging everything they owned in order to compete with large traditional businesses. The franchisees' work ethic and economic awareness taught other people to prospect in their towns for unserved economic niches. They showed the discrepancy in awareness between large corporations and consumers. But, lest we forget, there have been many franchise failures, in which both parties lost everything. (1) Where now are Shakey's and Taco Tico? (2) What happened to Arthur Treacher's Fish and Chips? (3) Whither Red Barn? (4) Are there any Spudnuts franchises left? (For answers, see the bottom of the page.)*

By 1970 there were 1,200 franchisers in the United States with 600,000 franchisees, doing $100 billion in sales. In the 1990s franchising accounted for about 35 percent of all U.S. retail sales, nearly $1 trillion a year. Recently franchising has converged with "easy money." After years of resistance, McDonald's began to accept credit and debit cards in 2004, and to its surprise, the average sale jumped from $4.50 to $7.00.

Franchising went overseas early. Hertz Car Rental and American Express were in Europe in the late 1950s. Manpower Incorporated had 130 offices in Asia and Latin America in the 1960s. But it was tough going. "You are not competing against amateurs," said Manpower president Elmer Winter, "but against some of the largest and most skillful merchants in the world."[15] Everything had to be rethought: "Let Hertz put you in the driver's seat" could not be translated as it first was: "Laissez Hertz vous faites un chauffeur" (Let Hertz make you a chauffeur). Most U.S. franchises never caught on overseas. Car washes, beauty salons, haircutting emporia, carpet cleaning, and donut shops were outside foreign cultural practices. By the mid-1990s, however, around 350 U.S. franchisers had 31,000 franchisees outside the United States.

Among the early successes was Domino's, with 450 restaurants in twenty-five countries by 1991 and overseas sales of $150 million. But it had to overcome numerous hurdles. Not every culture eats cheese. Nor is the idea of "home delivery" understood or feasible: in Japan, there are no street addresses. Some countries do not have laws protecting the names and logos of franchisees, so imitators spring up instantly. Other nations— Taiwan and Hong Kong are surprising examples—have rules preventing the repatriation of profit and arbitrary limits on franchisee fees. Foreign governments often treat franchises as special cases requiring extra fees, licenses, taxes, or building codes, and ensuring franchisee standards is

*(1) Shakey's and Taco Tico went out of business. (2) Arthur Treacher's is mostly limited to Florida, the New York City area, and a few other East Coast areas today. (3) The last Red Barn died in 1986. (4) There are two Spudnuts (doughnuts made from potato flour) franchises left, one in Berea, Ohio, and one in Richland, Washington.

always a problem for international franchisers. McDonald's works fanatically to make its French fries taste the same worldwide, and it has sued to revoke the license of its largest franchisee in France because he would not keep his fourteen stores up to its standards.

Franchising's ability to turn average people into economic prospectors is legendary among insiders. Den Fujita, the "Makudo King" of Japan, is well known. But more common are smaller success stories, like that of Roxana Orellana of Payless Shoes, recounted in the *Wall Street Journal*. Leaving war-torn El Salvador at fifteen, smuggled across the Tijuana border in the trunk of a car in 1983, she lived with a sister and supported herself by house cleaning while attending high school in Ontario, California. She started at Payless as a part-time clerk in 1993 for five dollars an hour. Three promotions and three stores later, she became a manager making $33,000 a year and buying company stock. When Payless noticed that Latinos had been sending Payless shoes back to Central America, it asked for volunteers to open stores in countries like El Salvador. Ms. Orellana volunteered, and now she runs her own store there.

Although franchising developed in the United States, it is now an export of several nations, most prominently Australia. With a population of 20 million, Australia has eight hundred homegrown franchise operations, four times as many per capita as the United States. Australians have used their cultural affinities to move into the U.S. market. For example, two-thirds of the households in both nations have pets; Aussie Pet Mobile, which washes and grooms Fido, has 205 U.S. franchisees. Bark Busters, which stops dogs from misbehaving, is another Australian franchise, with 52 American franchises. Computer Troubleshooter has 141 franchises in the United States. The latest Aussie franchise is Cartridge World, which refills our printers' ink tanks at 50 locations and was selling 5 new U.S. franchises a week in late 2004. "We want to be the McDonald's of ink and toner," said its CEO.[16] Our common language makes Australian franchises invisible, but another reason Americans see fewer foreign franchises is that U.S. franchisers are alert to invaders. When Mexico's El Pollo Loco began to open franchises in the United States in 1980, Denny's promptly bought all U.S. rights for $12.3 million. However, Stanley Kaplan Learning Centers, a subsidiary of the Washington Post, hasn't been able to slow the progress of Japan's Kumon Math and Reading Centers, which has 1,180 franchises in the United States (and 22,000 worldwide). Tutoring is a $4 billion a year business in the United States, and getting kids prepped for school is something the Japanese know about. It recently expanded into "Kumon Junior" for kids three to six, which is wildly popular with Asian American parents. It expects to grow 12 to 14 percent a year for the next decade.

Franchising has long financial coattails in globalization. Because they depend so much on leasing, franchisers and franchisees overseas have lobbied for and won changes in local tax laws and accounting standards. Leasing allows companies to invest less in fixed assets like buildings and vehicles, and to invest more in growing their businesses, which may mean advertising, research, hiring people, or buying rights. Payless Shoes' headquarters, for example, hired Mexican talk show host and *fashionista* Ana Maria Conseco to stump for its pumps in Ms. Orellana's El Salvador. The leasing of land and buildings, ad campaigns with spokespersons and tie-ins, and an orientation toward the mobility of consumers are predictable follow-ons. But the most pervasive influence of franchising may be invisible: that middle-management level of "consultant," the orchestration of the franchisee to the overall plan of the franchiser, the imposition of uniform standards, the emphasis on cleanliness, the training of franchisees, and the systemization of training for employees. This level of business sophistication is new in the underdeveloped world. In Southeast Asia, KFC and Wendy's operate "universities" that provide the layman's equivalent of an MBA. In some countries, this is the best business education available.

Airfreight

After the attack on the World Trade Center on September 11, 2001, international air traffic came to a halt for ten days. That seemed logical if you were a passenger or a politician, but there were protests from unexpected sectors—hospitals, high-tech manufacturers, and even whole nations. Japan imports 60 percent of its food and 30 percent of its medicine, and it was running out of essentials after the first week. Its just-in-time manufacturing systems lacked parts from Southeast Asia, so hundreds of factories shut down. As for luxuries, California strawberries had disappeared and Boston tuna more than doubled in price.

In Taiwan, the halt forced EVA Airways Corporation to cancel its thirty-one weekly flights to the United States, which would have been filled with Dell computers. But those computers were not built, since Dell's flexible manufacturers ran out of Intel's Pentium CPUs. Dozens of Taiwanese factories were poised to shut down, laying off thousands of workers. At Hong Kong International Airport, seven hundred tons of cargo bound for the

United States overflowed from warehouses. Planes loaded with clothes, toys, and electronics unloaded on the tarmac. Freight-forwarding agents had rented every free foot of depot on Hong Kong's islands.

Airfreight—an almost invisible element of globalization—was suddenly in the spotlight. The pressure to resolve air security issues and to reopen airports came less from passengers than from thousands of companies, which in a few more airfreight-less days would have gone bankrupt. Airfreight is as much an invention of logistics as of technology. Airplanes are the visible part, but the logistics of moving cargo quickly and profitably are a very sophisticated planning exercise. Everything from horses to human hearts to rice for Afghanistan moves by airfreight. We don't see it, because most airfreight arrives at and departs from special terminals, usually at night. There are whole airports devoted to airfreight, such as the Airborne Express operation outside Columbus, Ohio, opened in 1979. Airfreight also flies on passenger flights, but it is loaded first and unloaded last.

Airfreight is a collection of systems—computing, satellite navigation, trucking, parcel tracking, airplane leasing, advertising, aviation logistics, and materials science—that originated in the United States and took fifty years to achieve its present form. The United States remains the world's dominant airfreight provider, in part because it is the world's leading importer, but also because of the logistical complexity of the business. If airfreight were just a technology, it would be easy to copy. But it seems to require a "logistics culture."

The U.S. Post Office provided, by most accounts, the stimulus for the first airfreight enterprises. By the mid-1920s it was sending 14 million letters on planes flying 2.5 million miles a year. One carrier was Ford Motor Company, which won the contract to carry mail between Detroit, Chicago, and Cleveland, using the same planes on which it shipped parts between its factories. Its new sideline moved Ford to produce a plane for the job, the Ford Trimotor, in 1927. Made of duraluminum, twice as strong as aluminum, with three engines, the Trimotor could fly higher and faster, up to 130 miles an hour, than any comparable plane. Nicknamed the "Tin Goose," it also had twelve passenger seats, a cabin high enough to walk through, and work space for an attendant (the first were nurses, to deal with airsick passengers). Ticketed passengers were an offshoot of airfreight.

The U.S. government fueled airfreight's growth with the Contract Air Mail Act of 1925 and the Air Commerce Act of 1926. The first put airmail routes up for bid, while the second provided for designation of air routes, navigation systems, pilot and aircraft licensing, accident investigation, and other support systems. Winners of the initial contracts were firms such as National Air Transport and Varney Air Lines (both absorbed by United Air Lines) and Western Air Express (eventually part of American Airlines). But Juan Trippe, one of the original partners in little Colonial Air Transport, which carried the U.S. mail from Key West to Havana, later on founded Pan Am.

Despite airmail and the Trimotor, there were only three hundred aircraft in U.S. airfreight by 1939, when Germany marched into Poland. An aircraft construction frenzy began, producing fifty thousand planes a year by 1945. Most were fighters and bombers, but some were for freight, refueling operations, and troop transport. Technological innovations were built into these planes, but the logistical knowledge gained during the war was more important. How to load a plane for bad weather, how to improve fuel economy, how to avoid "deadheading" home—thousands of veterans returned home from the Air Transport Command and Naval Transport Service with knowledge of these complexities. Buying a couple of surplus airplanes, usually propeller-driven DC-3s, the vets started "GI airlines" or "non-skeds." Headquartered in garages in Hawaii and Los Angeles, operating in remote hangars, the non-skeds would fly anything to anywhere. Competition was severe, but the best-organized survived, firms such as Flying Tiger and Seaboard World Airlines. To fly "anything anywhere," they needed returning cargo, contracts for fuel and service, and local pilots familiar with weather, language, and airfields.

As Richard Malkin has noted, the 1950s were an era of rapid development and shakeout in this field, like the 1990s for the Internet. Traditional shipping agents entered the airfreight forwarding business. A division grew between the small package and air express market and the less time-sensitive freight business. Regular airlines in the United States and overseas jumped into both businesses. Who was legally permitted to do what, when, and where was a question that went to court. The Civil Aeronautics Board established systems of certification that identified certain airlines and forwarders as "direct" or "indirect" carriers.

More military improvements arrived, notably the Boeing KC-135, a refueling tanker that was revamped and introduced in 1958 as the Boeing 707, the first U.S. passenger jet. Jet engines were not only more reliable than piston-driven ones, producing less stress on the airframe, but they burned kerosene that was half as expensive as the hi-test gasoline required by piston engines. Other developments were on the ground. A Harvard study, mostly financed by Emery Air Freight, showed the cost-effectiveness of airfreight. Reducing delivery time, it argued, saved in production, distribution, and marketing costs. There was a whiff of the Japanese *kanban* system about this view. Emery printed do-it-yourself forms on which customers filled in blanks and were led to calculate the savings offered by airfreight. Whether accurate or not, the study attracted customers. The freight departments of many airlines began to rival the passenger divisions in revenues, increasing their corporate clout. But instead of breakout growth, airfreight remained a toothless tiger, because ground handling was still expensive. Remember Don Wetzel, the ATM inventor? He was working on automated baggage handling equipment in 1968.

Automated freight handling did finally appear, but early systems were slow and unable to deal with odd-sized packages. The key was to replicate somehow the system of containerized freight used in ocean shipping. So initially airfreight used wooden pallets, like the trucking industry. But pallets had to be secured to the fuselage or they would shift in flight, and on the ground they could not be broken open to remove some packages without restrapping the entire load, an awkward and time-consuming operation. Then special lightweight airfreight containers were invented that fit into slots in the plane's body. The distribution of weight inside the plane was reengineered, so that it accommodated a mix of containers and pallets but required less handling.

The deregulation of the U.S. air industry in 1977 led to the birth of freight "integrators," such as United Parcel Service and DHL. These companies not only offered airfreight but assumed the roles of collection and delivery as well. They expanded into trucking and invested unheard of sums in computers to handle logistics. Most radically, they built their own airports and terminals.

If we had to pick the moment that finally signaled the arrival of modern airfreight, it would be April 17, 1973. On that day a start-up company called Federal Express delivered 186 packages to twenty-five U.S. cities, from Rochester, New York, to Miami, Florida. It flew fourteen small airplanes out of Memphis International Airport, which had extra hangar space and was the center of FedEx's targeted service area. Memphis had been chosen after studies showed that it almost never closed due to weather. That kind of rationality typified the founder, Frederick W. Smith, who didn't come out of an aviation background at all. As a Yale student in 1965 Smith had written a term paper on the inadequacies of airfreight. What was needed, he averred with undergrad gusto, was a special system for moving medicines, computer parts, and electronic gear. His FedEx did not make money until 1975, but after deregulation it found investors and leased Boeing 727s and McDonnell Douglas DC-10s. Smith advertised early and heavily, spending $150,000 in his second year alone. The effort worked, as parcels rose from 3,000 a day to 10,000 after the campaign. By November 1988 FedEx was handling 1 million packages a year. Its growth since then has paralleled the rise of just-in-time manufacturing systems:

1984   Offers service to Europe and Asia
1985   Offers direct, regularly scheduled service to Europe
1988   Offers direct, regularly scheduled service to Japan
1989   Purchases Flying Tiger and expands to twenty-one new nations in South America and Africa
1995   Purchases Evergreen International, then the only U.S. cargo carrier with rights in China
2000   Handles 3.3 million packages, can transport 26.5 million pounds, and can fly 500,000 miles in one day with the world's largest all-cargo air fleet
2004   Achieves greater gross transport capacity than the combined airborne components of the U.S. military

Today Americans receive countless small packages from FedEx and UPS, its much bigger brother, but they don't realize how many heavier items move by airfreight. Medical equipment, many fruits and vegetables, commercial flowers, exotic birds, tropical fish, thousands of daily news-papers and magazines and journals, elephants and giraffes for zoos, and whole tuna destined for Tokyo's Tsukiji fish market are all shipped by air-freight. Magic Millions–IRT will pick up your horses at the barn and air-freight them anywhere in the world on its specially equipped Boeing 747s. Whole U.S. communities owe their existence to airfreight: the ferns-for-florists capital of North America—Apopka, Florida—ships $85.5 million worth of the green stuff a year, and all orders to Europe and Japan, by air-freight. France airfreights 100,000 bottles of *Beaujolais Nouveau* to the United States and Japan every November. Precision equipment, from computers to milling machines to cyclotron components, travels by air-freight because there are fewer shocks en route. Anyone buying any kind of computer can be sure that parts, if not the computer itself, traveled by airfreight. The electronics, the airbags, and the emissions control systems of United States and foreign autos built in Mexico are flown in daily from factories around the world.

The range of mundane merchandise now flown by airfreight just be-cause purchasers want it quickly is astonishing. The CAB has fined airlines for violating safety guidelines for flying paint, solvents, and other inexpensive, nonperishable products. FedEx made the news in early 2002 when it failed to detect leakage from a large, heavy box of mildly radioactive materials. That was hospital garbage flying airfreight.

Having an airfreight facility has become critically important for underdeveloped nations that want to export. Taiwan built several in the 1980s, and Malaysia helped itself greatly by establishing several air cargo terminals in the 1990s when it was flush with oil money. These assure Dell or Siemens or Fujitsu that computers or other products built in these countries leave quickly and securely, and that critical parts to build them can be supplied overnight. In booming China, the roads and railroads are so poor that airfreight is on its way to becoming the de facto shipping method. After China Eastern Airlines began twice-weekly airfreight flights from Shanghai to Los Angeles in 1999, Shanghai's exports shot up. UPS airfreight out of China in 2004 rose 70 percent over 2003, while that of FedEx increased 50 percent in the same period. A newcomer, Polar Air Cargo, won nine new weekly flights in and out of this dynamic market. Cathay Pacific (Hong Kong) has an astonishing twenty-one cargo flights a week to Dubai. In most nations that make shoes or clothing, the raw materials are flown in and the finished products flown out. Most suppliers dealing with Wal-Mart, for example, end up using air cargo for at least some products.

Other nations besides China also look as if they may skip, or scant, traditional road-building in favor of airfreight terminals. Airports are cheaper than railroads, and the post 9/11 world is overstocked with airplanes. In the U.S. airfreight was a $48 billion business in 2004. But another figure says it all. According to Hendrik Tennekes, by 1996 the cost of flying one ton one mile had dropped to fifty cents. Bert Hall of the University of Toronto claims that this has been "the most powerful influence for globalization ever."[17] But there is a shipping mode still cheaper.

Containerized Freight

The world's largest enclosed building is not the Pentagon, not The Mall of America, and certainly not a football stadium. Neither spectators nor shoppers ever see it. It is the Hong Kong Container Freight Station, with more than 7 million square feet of storage, serving the world's busiest container port.

On January 23, 2002, an event thousands of miles away—the bankruptcy of the Kmart Corporation—had a profound effect on this building. Millions of dollars worth of goods that Kmart had ordered—but not paid for—sat in the warehouse. Three Hong Kong companies faced a tough decision: Texwinca Holdings makes apparel for Kmart, Climax International makes stationery, and Boyo International makes artificial Christmas trees (a load of trees for the *next* holiday eleven months off was about to go out).

Should they ship the order and hope to be paid during Kmart's bankruptcy hearings? Smaller companies all over south China were in a similar quandary. "I don't believe Kmart will collapse, given its huge scale of operation," said Boyo deputy director Phillip Lam Pak-kin. "We are already working on next year's order. We hope sales from Kmart will reach 25 percent of our turnover."[18] If the companies stopped shipment, it would be very hard to find other buyers for the goods.

As it turned out, the decision was made for them and for Kmart by the world's container freight shippers. For more than twenty years, Maersk Sealand, American President Lines, and Overseas Container Lines had been shipping 40 percent of Kmart's cargo "free on board," meaning that no payment was collected until the goods reached their destinations. They decided that enough was enough—the Christmas trees would not load. Suddenly the reorganization and bankruptcy hearings picked up speed in the United States, where no one wanted Kmart to sink.

Containerized freight is another invisible technology, largely invented and popularized in the United States, that is propelling globalization. "Simply put," writes Toby B. Gooley, "containerization has raised the standard of living worldwide. Without containers, the prices we pay for consumer goods would be significantly higher—or the goods might not be available at all. Without containerization, the globalization of manufacturing could never have happened. Asia would not be the economic giant that it is today."[19]

The United States had the first and largest container ports. When the SS *Elizabethport,* then the world's largest freighter, arrived in Oakland's new container facility in 1972, that port was the second busiest in tonnage, and the second largest in acreage (after New York) in the world. The Sea-Land Corporation had modified 5,000 truck trailers at a cost of $600,000 to make them usable as containers. Ships like the *Elizabethport* could be unloaded in one-sixth the previous time. In the same year, Sea-Land opened a 232-acre private container terminal in Elizabeth, New Jersey, and the New York–New Jersey Port Authority was inaugurated. The New York Port Authority was the busiest in the world through 1985. The United States dominated this new field.

But the rest of the world has caught up. If current trends continue, the United States will choke on its inability to offload the goods it buys abroad. As the Kmart case shows, the globalization of container freight allows other countries, and even companies, to apply the same financial rules to U.S. companies that the latter have long imposed on them.

Today most containers cross the Pacific to the West Coast, where 65 percent of Asian freight arrives, 11 million containers a year at Los Angeles and Long Beach. New York is now the 13th largest container port, receiving 4 million containers a year. From the West Coast, the goods are trucked across the United States. The ships that do make it to New York, like the new *Hyundai Glory,* are so big that they squeeze through the Panama Canal with inches on either side and have to wait for high tide to clear lower New York Harbor and dock. They carry containers stacked thirteen high and thirteen wide—2,087 of them at one time. But when they leave New York, they are basically "dead-heading" back to Asia, carrying as few as 300 containers of cargo and 1,000 empties. Most of the filled containers are stuffed with New York's biggest exports: waste paper and scrap metal. And the new larger freighters now being built won't call on New York at all until a deeper $2.2 billion channel is completed there a decade from now.

Defining the international standard "container" was the work of good, gray engineers and bureaucrats. In the United States it became the hobbyhorse of a retired Alcoa engineer named Herbert Hall and an International Standards Organization (ISO) official named Fred Muller, who enlisted a young ISO engineer named Vince Grey. There were comparable groups in other nations, but the United States was the prime mover. As Grey recalls, the crux of the standardization problem back then lay in railroads.[20] The container was in effect a railroad car that sat on a chassis from which it could be removed. It had to absorb all the shocks normally delivered to a railroad car, including coupling and uncoupling when trains were made up. This stress was compounded by the length of the train. The last car added to a train is essentially running into an immovable wall. And the first cars behind the engine, when a train departs, are momentarily pulling the weight of hundreds of cars, a stress of millions of pounds. How many times could a container be subjected to such stresses without failing? That depended on the standard to which it was built—and every country had a different standard, which affected all the rolling stock of its railroads. Every nation wanted to have *its* standard adopted internationally to avoid the expense of building new railcars. The initial U.S. proposal was rejected, Grey recalled, because there were different national standards for speed and impact in making up trains in different countries.

It required years of study and negotiation before the ISO approached the definition of a container. That happened at a meeting in Moscow in 1967, but only when executives of major U.S. shipping companies intervened personally. Suddenly, Grey recalls, "There were these executive vice-presidents of these major companies, and they sat at a drafting board and drew these drawings." There emerged ISO/TC 104, which defined an ISO container. The degree to which American needs drove the container standard is illustrated by the standard container sizes: for example, 10 feet by 8 feet by 6½ feet, 40 feet by 8 feet by 6½ feet. The metric equivalents (6.06 by 2.44 by 2.59 meters, etc.) are used outside of the United States, of course, but even the newer variants—the 40-foot "cube" and 45-foot "high cube"—are advertised and rented worldwide by their American measurements and nicknames.

Although many subtypes of container have appeared and some national variants exist, international shipping containers are basically the same. They ride on every railroad and expressway, and are stacked in every port in the world. They come in heated, refrigerated, airtight, and liquid versions. Corporations exist simply to own containers or to lease them. Decommissioned containers, with windows and doors cut into them, serve as housing for the poor in South Korea and China. They serve as construction offices and work sheds in Japan. Today China is the leading manufacturer of shipping containers, followed by South Korea.

The development of the freight container in the United States was doubtless prompted in part by the shenanigans of the International Longshoremen's Union (ILU) in the United States. The ILU went on strike in 1952, 1954, 1957, 1959, 1962, 1964, 1968, 1971, and 1977. Most of these strikes lasted more than fifty days, causing presidents from Truman through Nixon to invoke the Taft-Hartley Act seven times. The ILU resisted all forms of mechanization beyond the forklift, established high wages for unskilled labor, closed down ports during disputes, pilfered cargo or, in some notorious locales, allowed it to be hijacked, and created "make work" that was legendary in labor management circles. It became twice as expensive to load a boat in New York as in Houston. The situation attracted congressional investigation and even film makers (Elia Kazan, *On the Waterfront,* 1954). When shippers finally won the right to use containers in 1969, there was still a "50 Mile Rule" that prohibited them from delivering freight within that radius of the port—to preserve other union jobs in trucking. East Coast longshoremen finally killed their own jobs, as shipping moved to the Gulf Coast.

In 1980 an industry-labor Container Carriers Council paved the way for container freight nationwide in the United States. Next up was re-thinking the dock. Too big for forklifts, containers required cranes. Mechanical unloaders for coal and ore, called Huletts, had long been used on the Great Lakes, so the idea of dockside mechanization was not new. In the new U.S. technology, huge cranes straddled the ship, lifting containers from the hold and placing them directly on trains and trucks. Cranes could also be mounted on ships, further reducing the need for shoreside assistance. Gang size, reduced by five after the 1980 agreement, dropped again, but there was still too much labor involved, because the container's contents had to be verified, customs inspections passed, and duties paid. Some of this was automated by hand-held computers, which French longshoremen continued to call *les marteaux*. But many functions had to be repeated each time a container passed from boat to rail to truck. The solution to this paperwork, at least until September 11, 2001, was the uniform TARE container sticker, a variant of the bar code. The confluence of these technologies resulted in "intermodal freight," which used the *same container* on ships, trains, and trucks. But to do this, the truck trailer had to be modified. At first containers were dropped on to a bare truck chassis, but then modified chassis were built into containers.

By 1995 a ton of shrimp caught in the Indian Ocean could be frozen, packed in a freezer container in Bangkok, transported by ship to Los Angeles, by train to Chicago, and by truck to Champaign-Urbana, without spoiling and usually without being opened. The savings in labor costs was enormous—the labor cost alone had made the process unthinkable ten years earlier.

The impact of containers on globalization is inestimable. Not only artificial Christmas trees from China and shrimp from the Indian Ocean, but cloth from India, clothes from Turkey, furniture from Bali, wine from France, skis from Austria, olive oil and books from Italy, beef from Australia, flour and soybeans and precision parts from the United States, machine tools from Germany—millions of products ship in containers.

The United States led the world to container shipping, but other nations have caught up. Mitsui O.S.K. of Japan, for example, operates 350 container ships, far more than any U.S. company. Maersk Sealand of Denmark owns 800,000 containers, for use on its 250 ships, including the world's largest boat, the *Sovereign Maersk*. More than 350 yards long, this ship carries 6,600 containers and cruises at twenty-five knots an hour.[21] Most container freight today is carried on Asian-owned ships, including the Hyundai and Hanjin lines in South Korea and the Yang Ming line in south China.

What happened to American supremacy? Other nations were more flexible, especially about land use, and are less "green." Take the case of Japan, which became a major shipping center for the United States during its occupation, the Korean War, and the war in Vietnam. As Alex Kerr notes in *Dogs and Demons*, Japan had a centuries-old tradition of dredging, filling, and harbor making. In 1967, while Americans scoffed that it would sink, Japan opened a completely man-made island in Tokyo Bay called the Shinagawa Terminal. More advanced than Oakland, this terminal was fed by truck-only roads and rail lines right to the docks. It had acres upon acres of warehouses. Shinagawa became the model for the developing world.

Today there are new state-of-the-art container ports in places that Americans have never heard of—Mindanao, Balawan, Cochin, and Port Rashid. Many were built with loans from the United States or World Bank. Their importance is understood by locals: when the Cochin container port in India's Kerala state opened in 1988, it was inaugurated by Mother Teresa.

In the United States, however, there has been great resistance to building new container ports or expanding existing ones. The Port of Long Beach, at 4.1 million TEUs (twenty-foot equivalent units, the basic measure of containers), has the greatest warehouse capacity of any U.S. container port, followed by the Port of Los Angeles. But together they have about half the TEU capacity of Singapore, and neither port has anywhere to expand. Environmentalists and residents of Los Angeles sued to block the opening of a $97 million port expansion finished in 2002. The big problem is the "greening" of American values. Container ships burn the dirtiest, cheapest grade of diesel fuel, and they run their engines in port to provide on-board electricity, making them a major source of pollution. L.A. already has severe pollution problems. The channels dredged for the freighters' passage stir up toxic chemicals. But the Los Angeles port opponents' biggest beef is that tall new cranes spoil their ocean views. In Long Beach, opponents want a twenty-foot-high berm to hide the port altogether. Americans want cheap goods from abroad, but they don't want to see (or smell) them arrive.

Other U.S ports are similarly beleaguered. According to the *Wall Street Journal*, more foreign tonnage was landing at Houston by 2004 than at any West Coast port, so Houston's port authority began to turn away shipments, particularly from Asia. A new $1.2 billion port was announced in 1998, to be built on untouched land farther from town, but it has been blocked by environmental concerns and local residents. "You can't count on Houston to work the ship in timely fashion," one shipping executive now says.[22] Typically it takes twenty-four hours to unload one of his ships, but it's eight hours longer in Houston—which adds $5,000 to $10,000 to shipping costs. On the East Coast, South Carolina residents have blocked expansion that would add eight hundred acres to the port at Charleston, presently the second largest on the East Coast. Local environmentalists also oppose a port at Savannah, Georgia, which Wal-Mart would like to turn into one of its East Coast import hubs. On the Great Lakes, a proposal by CSX Intermodal to build an 850-acre container terminal on the site of an existing junction yard in Detroit met stiff opposition, because it required the demolition of eighty-one homes and ninety-nine businesses. Built up in the 1920s and 1930s, the formerly Polish neighborhood now houses Yemenis, Romanians, and Mexicans, just the immigrant salad that the United States should be proud of, opponents say.

But new terminals have to be bigger, says Robert Haessler of the University of Michigan, to handle the latest advance in containerized freight.[23] That is the "double stack," in which trains carry stacked containers. Every train thus loaded, Haessler argues, takes one hundred trucks off the road. If Detroit stands pat, it will no longer be competitive with Los Angeles and New York in manufacturing, says the mayor's office. That means more jobs will disappear.

But American container ports are already not competitive with ports at Gioa Tauro, Port Kelang, and Keelung—which the Motor City has never even heard of. In 2001 the small Vietnamese city of Binh Duong up and built a state-of-the-art six-thousand-square-meter container port. With seven hundred meters of quayside dock and six cranes, the port only handles ships up to ten thousand tons—not as big as Detroit, but more efficient. With a sister development at Cat Lai accepting ships up to forty thousand tons, little Binh Duong will battle Detroit for jobs as soon as global industries set up shop there. Binh Duong can make bumpers and brake shoes just as well as Hamtramck can. It is not hard to see that when goods move by sea, underdeveloped nations have logistical advantages by virtue of later starts. They are more flexible in land use, and their citizens are not as politically empowered as Americans to prevent pollution.

On the island of Nusa Lembogan, in Indonesia, I saw a new local industry in 2002 that container shipping had made possible. Hundreds of islanders, who might otherwise be scrabbling for the barest living, were instead farming a special seaweed that was used for expensive European cosmetics and as an organic food additive. Grown on underwater trellises over the coral reefs, this crop was harvested at low tide and then dried. But it could only be dried a certain time without losing desirable properties, so every week a small boat arrived (the island had no dock, much less a port) to collect the seaweed and take it to Denpasar, where it was packed in a temperature-controlled container and shipped to Rotterdam.

As the World Bank has pointed out in several studies, even though container ports have stimulated growth and employment in underdeveloped nations, raising the standard of living, they have unforeseen consequences. One is that walking, still the most common form of transport in the underdeveloped world, gets slighted. The rush to create container facilities and airfreight terminals has meant less rural road construction. Walking, bicycling, and rickshaws require roads. Without them, it is difficult to attend schools, reach medical clinics, or get goods to local markets, particularly in Africa. Will these technologies spread the American disinclination to walk? Further, when local roads are built, construction is often too shoddy to resist the weather. Maintenance of container ports looks relatively easy compared with the maintenance of rural roads.

With more than 20 million containers now in use worldwide, security has also become a concern. On the mundane side, manufacturers in the age of lean production want to know precisely where their supplies are. And retailers want to know where the manufacturers' shipments are. But after the attacks on the World Trade Center, the possibility of using a container as a weapon is taken seriously. Twelve million containers enter the United States every year, many from Dubai and Algeciris. The state-of-the-art of container verification is relatively poor: at present about 5 percent of containers every year are misrouted, stolen, damaged, or even lost. And these are very large, not to mention expensive, objects to lose.

But there may be a high-tech American answer. "Smart tags," and even more expensive sensors capable of monitoring temperature, radiation, motion, and tampering, will soon be part of many containers, adding $50 to $150 to the $1,800 cost of a container (or $600 for sophisticated models). Secure bolting systems will replace the plastic strips often used for closure. The U.S. Department of Defense already has 25,000 containers with radio transmitters and receivers moving around the world every day. After 9/11, the United States began to open and inspect a higher percentage of the containers that enter U.S. ports. In 2004 it proposed a "green lane" at port customs for containers equipped with smart tags or RFID sensors, allowing them to avoid further inspections. The volume and importance of U.S. trade is such that these new containers will be in place worldwide within a few years. And already numerous small companies in the United States are gearing up to meet the demand.

Bar Codes

In 1948 the president of a grocery store chain walked down the hall with a dean at Philadelphia's Drexel Institute of Technology. The executive had given money to the college and, in the way of donors, he explained that he had a problem he'd like the school to work on: tracking inventory in his grocery stores automatically. The dean rebuffed the donor. But walking behind them was Bernard Silver, a graduate student. He overheard, and his mind lit up. Silver took his idea to a friend, a twenty-seven-year-old graduate student and university instructor named Joseph Woodland.

Woodland thought they might "use patterns of ink that would glow under ultraviolet light," reports Tony Seideman. "Silver and Woodland built a device to test the concept. It worked, but they encountered problems ranging from ink instability to printing costs." Woodland became obsessed, quitting his job, selling some stocks, and moving to his grandfather's apartment in Florida to work full time on a solution. He thought he could blend Morse code with Lee De Forest's 1926 optical scanning technology for film sound. "I just extended the dots and dashes downwards," he said later, "and made narrow lines and wide lines out of them."[24] De Forest had recorded sound as degrees of transparency at the edge of the film, through which light shined—the rays that passed through were measured and translated into wave forms that could be amplified as sound. Woodland's system was even simpler, since he wanted only to measure thick and thin lines. To allow the code to be read from any direction, he formed it into a bull's-eye. He and Silver applied for a patent on October 20, 1949.

In 1951 they built a model in Woodland's living room. The machine was the size of a desk and wrapped in black oilcloth to keep ambient light out and some of the scorching heat inside. The intensity of their five-hundred-watt light bulb burned up the bull's-eye codes; in fact, it was so strong that it caused eye damage, and only a small portion of the light could be read by the photomultiplier 935 tube. But the system worked. To register the results, they needed some kind of electronic tabulating machine. But the only computers back then were huge, expensive, and required punch cards. So there the bar code scanner sat.

Woodland went to work at IBM in Binghamton, New York. He persuaded his employer to look at the bar code reader, but IBM declined to fund it, saying the necessary technology was ten years away. Instead, it offered to buy the patent for a low price. They got a better price from Philco, but Silver died unexpectedly in 1963. With most of the seventeen-year patent gone, Philco sold its rights to RCA.

Now imagine a business completely different from selling groceries—running a railroad. Usually railroads lent out their cars and borrowed those of other companies with only a foggy notion of where their property was. David J. Collins, an MIT undergraduate, worked summers at the Pennsylvania Railroad. After taking his M.S. degree in 1959, he signed on with the Sylvania Corporation, which hired him to find uses for a military application it had: the computer. Collins thought of railroad cars. He invented a system of orange and blue reflective stripes representing the numbers zero to nine. Each car had a four-digit number revealing the owner, and a six-digit stock number. As railcars rolled into a yard, a colored light would shine on a coded placard and be reflected by the orange stripes, but not by the blue ones. Boston & Maine Railroad tested the system on gravel cars in 1961 and by 1967 had the problems worked out. But Collins thought the invention had broader application. He told his bosses, "What we'd like to do now is develop the little black-and-white-line equivalent for conveyor control and for everything else that moves."[25] But the company said, "We don't want to invest further. We've got this big market and let's go and make money out of it."

Collins quit and cofounded Computer Identics Corporation. His new company prospered with his technology, but Collins was interested in the laser, just becoming affordable in the mid-1960s. A tiny laser beam did a better scanning job than his colored light or Woodland's five-hundred-watt bulb. Its thin beam was absorbed by black stripes and reflected by white ones, giving sensors a clear on or off signal. Furthermore, lasers could read codes from three inches to several feet away, and they were so fast that they could scan several times a second, helping to decipher dirty or torn labels.

In 1969 Collins installed the first modern bar code system. It went into a General Motors plant in Pontiac, Michigan, where it recorded the production of auto axle assemblies. Since there were only eighteen types of axles, the bar codes were simple. He installed another at General Trading

Company in Carlsbad, New Jersey, where it directed shipments to the proper truck-loading docks. There were fewer than one hundred options, so neither installation needed more than two digits of information. Collins made the units by hand, forming the housing by turning over a wastebasket and molding fiberglass around it.

RCA executives, meanwhile, attended a convention of the grocery industry, at which development of bar codes was enthusiastically supported. Wallace Flint, a grocer's son, had proposed such a coding system at Harvard in 1932, but he lacked the technology. Though fearful of antitrust prosecution, in 1969 the industry developed the Universal Grocery Products Identification Code (UGPIC). The Kroger chain in Ohio volunteered for the experiment, and RCA deployed a task force at its Princeton, New Jersey, lab to conquer the problems: the bar codes had to be readable from any angle and distance, they had to be cheap, and the systems had to pay for themselves within two or three years. If all that could be done, consultants predicted $150 million a year in savings, a considerable sum in those days. That was a gross underestimate. "It turned out there were massive savings that we called hard savings—out of pocket savings in labor and other areas," said Alan Haberman, chair of the industry committee. "These included checking out items at twice the speed of cashiers using traditional equipment."[26]

Unfortunately, what RCA brought back to the industry in 1971 was the bull's-eye bar code of Woodland and Silvers. It demonstrated this system at a grocery convention, where IBM executives in attendance noticed the commotion and remembered that Woodland was still on their staff. Kroger put the RCA system into operation in a Cincinnati store, but it had problems. Printing presses sometimes smeared ink due to inertial forces, so the bull's-eye did not scan properly. IBM called up Woodland and transferred him to North Carolina, where he worked with George Lauer, who developed a system of splitting the bar code into halves of six digits each.

Lauer's solution was key. The modern bar code begins with a "guard pattern," two taller pairs on the left that identify the beginning of a Universal Product Code. The next digit tells the type of product (for example, if it should be weighed); the next five are the manufacturer's code. Then there is a middle guard pattern, again taller. The first five pairs to the right are the product code; and the last is a check digit, used to verify that the preceding digits have been scanned correctly. Finally there is a right guard pattern. Verification is ensured in two ways. The sum of the bar widths on the left is odd. The sum of those on the right is even. The check digit is three times the sum of the digits in even positions plus digits in odd positions, which should equal a multiple of ten. If it doesn't, the code must be reread. Lauer invented the UPC in its current form in 1973. To compensate for printing processes, IBM made the bar code straight, so that extra ink left by printing ran harmlessly out the end of the code's lines.

IBM set up a system in Marsh's supermarket in Troy, Ohio. On June 26, 1974, a ten-pack of Wrigley's Juicy Fruit chewing gum was the first UPC-labeled item ever scanned. It is now on display at the Smithsonian Museum of American History.

Bar-coding spread very slowly. Only 1 percent of U.S. stores used it in 1978, a figure that rose to 10 percent by 1981. By 1984 the figure was 33 percent, and by the 1990s it was 60 percent. To be truly cost-effective, bar-coding required 85 percent of products to be labeled. Lauer worked on refinements of the system, and Woodland developed the European Article Numbering system (EAN), which added two digits.

Today there are variants named Code 39, Code 16K, and Interleaved 2 of 5, among many others. The EAN may soon replace the UPC as the most widely used system. Bar codes are affixed to railroad cars, shipping containers, automobiles, and even ships. The U.S. Army uses two-foot long bar codes on its boats at West Point. Hospital patients wear bar-coded bracelets. Biologists have attached bar codes to the backs of bees to study their mating habits. Hand-held bar code scanners use lasers that read codes from any angle, and at a distance, so that now one employee walking through a warehouse can take its inventory. Fed into computers using space-optimization software, such inventory can be compressed to take up the least room, a process known as "squeezing the air out." As the Japanese *kanban* systems showed, 30 to 40 percent of most warehouse space is empty.

One bar code innovation that most of us have seen is the FedEx codabar. This device scans shipping labels at each transfer, with data uploaded to central computers that permit customers to locate their shipments. When FedEx delivers a package in Johannesburg, the driver scans the label. The shipper, in Japan or Canada, then knows the package has arrived.

The technology is now worldwide, but we seldom think about its impact. In underdeveloped nations, bar-coding alters behavior. Everything from checks to candy to chairs used to disappear in shipment, whether sent by private carrier or the post office. The items "got lost." But today shipped items can't simply "disappear" once they are being tracked. Bar codes have become an enormous deterrent to theft during shipment. They also deter theft from stores and warehouses, because merchants know their inventory and sales more precisely. Bar codes allow small merchants to carry larger stocks. When I visited video rental stores in Amman, Jordan, for example, I found that they carried hundreds of tapes instead of the few I had expected. Owners told me that before bar code scanners arrived in 1999, they had been unable to track their stock, over-due dates, and the like, so they kept inventory small.

Bar codes invite merchants to enter a "supply chain," as Americans say, so they are not caught short of a particular item. Merchants who used to keep accounts by hand and estimate what and when they needed to reorder, now use bar-coding to carry more stock, and they get restocked more punctually and precisely. This is true not only of stores, but of doc-tors, pharmacies, and hospitals as well. Being in the supply chain makes critical and common services more dependable. Bar-coding has revolu-tionized the inventories of the underdeveloped world, especially in regard to spare parts and medicines.

Even as bar codes spread, however, more comprehensive product IDs are on the horizon, called radio frequency identification devices (RFIDs), mentioned in the preceding section. RFIDs allow a great deal of informa-tion to be broadcast from a tiny chip planted in an adhesive label. To be read, they need only be near receivers—so that dirt, light, and angle of reception do not matter. Newer tags, called Class 2 RFIDs, can also *record* information as they pass sensors, revealing where they have been, and when. In 2004 the U.S. Department of Defense and Wal-Mart announced that they would require RFIDs on a variety of items within two years. RFIDs are used on some highway tollbooth cards in the United States, but with costs between twenty cents and a dollar each, they cannot be slapped on everyday goods yet. That cost will drop sharply as the two U.S. giants push the new technology. And since Wal-Mart buys so much in China and other developing nations, RFIDs will appear in Asia more quickly than bar codes did.

Computing

To "google": the name of the search engine has become a verb for Amer-ican computer users. Created by Stanford University doctoral dropouts Sergey Brin and Larry Page, Google handled 138,000 queries a minute in 2004, not only in English but in eighty-nine other languages. In a typical day Google did 200 million searches of 6 billion web pages. Along with

Microsoft Windows, Google was the most widely used program in the world.

It is obvious that the worldwide computer revolution began in the United States, but that is too big a subject to tackle here. Besides, by 2000 many "American" computers were manufactured and even designed overseas, including those of industry leader Dell (Malaysia, Taiwan). Even Hewlett-Packard, long driven by its U.S. engineering school graduates, makes most of its machines in Asia, and in 2004 asked its Asian engineers to design and build servers in Asia for its Asian markets, because the components were mostly manufactured there. There was nothing particularly "American" anymore about computer hardware.

But computer software is a different story. Software implies a way of working, a set of habits, certain kinds of goals. Attempts to outsource the design of text editors to India, for example, failed because Indians didn't work the same way that Americans did. But the Internet itself, an outgrowth of the Defense Department's Arpanet, with its basics invented in 1973 by Vinton G. Cerf and Robert E. Kahn in a conference room of the Cabana Hyatt Hotel in Palo Alto, is pure Americana. Their paper "A Protocol for Packet Network Interconnection" described TCP/IP, the syntax of the Internet, and is so historic that a copy sold at auction at Christie's in February 2005 for several thousand dollars.

It may not be apparent at first how the Internet foments American habits abroad. By way of example, let's focus on the software that powers sites such as Google and eBay. These programs use artificial intelligence to interpret a user's searches, hypothesizing what she or he might be looking for or is poised to buy. Then there are portals such as MSN, Yahoo, and AOL, which know a considerable amount about their clients' favorite websites, e-mail contacts, and purchase patterns. These search habits and favorites are American. As these consumer-modeling programs are exported (or copied), they have the potential to induce shifts in work and purchasing patterns overseas.

If we are tempted to think software is not culturally inflected, we need only type "Moscow" into Yahoo's search engine. High on the results page is a link to Travelocity, one of Yahoo's partners. Type in "Moscow" at MSN in 2004, and we see a page of hotel guides that don't distinguish between Moscow, Idaho, and Moscow, Russia. Type in "Moscow" at Google and pictures of the city appear first, with a tourist guide later. As *Newsweek* demonstrated, entering "insomnia" into these three search engines produced equally disparate results. Google offered thirty suggestions for a better night's sleep, whereas MSN returned a movie starring Al Pacino and Robin Williams. Yahoo returned the movie plus commercial links to natural remedies for insomnia and ambient noise CDs. Software is culture-ware. These three search engines (plus AOL, Ask Jeeves, and a few others) dominate the rapidly expanding international market for search services.

Language does pose some resistance to their spread, especially in Asia. So do local cultures and patterns of computer use. Jean-Noël Jeanneney, president of France's national library, announced a $128 million plan in May 2005 to combat Google's plan to scan the holdings of several American libraries. But these resistances tend to succumb more quickly than anticipated to the very habits encouraged by software. Google is already the search engine of choice for 53 percent of all French searches, compared with 13 percent for France's Wanadoo, the national leader. And 250,000 of the volumes Google will scan at the University of Michigan are French or German.

With their unique, non-Roman syllabaries, however, Asian languages would seem to pose a problem for language-based query software. Keyboards are mostly QWERTY and use the Roman alphabet, but there are Asian versions. In the complex Japanese keyboard, which displays four syllabaries, most users learn to type in *romanji* (English) and *hiragana*, and then to click on the correct *kanji* (Chinese) character from a drop-down list. Users can search the Internet by these thousands of *kanji* or by their *hiragana*, *katakana*, or *romanji* equivalents. There are comparable difficulties in Chinese. It would seem to add insuperable difficulties to achieving the kind of success Google has achieved in the United States. But such seeming is often vanquished.

At the basic level, all languages are now reducible to "bits" in computers, and at the program level their "words" are simply "strings" of bits, which are numerically or logically related. With the adoption of Unicode in 1991–92, driven by American universities, not only *Han* (Chinese characters, Japan's *kanji*) but also Arabic ligatures and Vietnamese set phrases entered computing's basic bit-level repertoire. The constantly doubling speed of CPUs has rendered "translation" time negligible. By 2004 Unicode was available everywhere, as extensions to the Apple and Windows operating systems. So formerly difficult languages became, at least potentially, searchable by programs such as Google.

But the cultural categories of returned results may still differ, as they do in examples from the American programs. For example, entering the Japanese words for "water" (*mizu, yu, o-yu*) at Google USA, produces a set of hits for bath design and computer games ("mizu"), a set of hits for

Yugoslavia (where the national suffix is ".yu"), and a 26,500-foot peak in Tibet (Cho *Oyu*). But entering the same words in *kanji* at Google Japan produces far different results—more than 10,000 of them.

There are already language-specific, non-English Internet search engines. In China, which in late 2004 surpassed Japan as the nation with the world's second-largest total Internet population, three domestic companies were doing battle with the Chinese version of Google. Baidu.com is styled after Google and had an index of 220 million web pages in March 2004, scheduled to grow to 300 million "in the next few months."[27] A second company, 3721.com, was bought by Yahoo in November 2003 for $120 million cash. At about the same time, Yahoo United States announced that it was dropping Google and developing its own search engine. But as 3721.com founder Zhou Hongyi pointed out, it was taking extraordinary efforts to meld his technology with that of Yahoo, presumably because of language and artificial intelligence assumptions. The third domestic search engine, Zhongsou (China Search), had the largest index in March 2004—280 million websites—and was the search agent for such major Chinese portals as Sina.com, Sohu.com, and NetEase.com. Since then, however, Sohu dropped Zhongsou to develop its own search engine. Baidu, which made its first profit in 2003, was said to be a takeover target of Google. In a possible feint, Google introduced its own Chinese search engine in February 2004. Whether it grows through imitation, competition, or acquisition, Google will have a significant impact on China, which will become by 2010 the world's largest Internet user base. The way that Google's rules drive its searches is Google's gold mine, and those rules derive from American cultural context. Might there be Chinese rules? Or will the world search as America searches?

Amazon and eBay are other American companies with commanding positions on the Internet outside the United States. Few would deny that their operating procedures are inherently "American." From the "shopping cart" to their seller-buyer rating systems to their laissez-faire attitude toward sales tax, they epitomize the free market. EBay sells everything from Florida condos to secondhand clothes and sexual aids. Not only are sales basically private, but dispute resolution has been privatized. Not since the heyday of the British East India Company or the Union Pacific Railroad has a company so dominated a swath of business. As of early 2004, there were eBay sites in twenty-two other nations. According to CEO Margaret Whitman, eBay's payment subsidiary Paypal reported that 10 to 12 percent of its payments were international money transfers, eliminating money orders and currency conversion fees.[28] In fact, eBay has become a bank, an international one. The nations it served in 2004 were:

North America: United States, Canada, Mexico
Europe: Austria, Belgium, France, Germany, Great Britain, Ireland, Italy, Netherlands, Spain, and Switzerland
South America: Argentina, Brazil
Asia: Australia, China, Japan, Hong Kong, Korea, New Zealand, Singapore, Taiwan

This list tells us that if a nation doesn't have reliable Internet infrastructure, dependable delivery services, secure international banking, or laws that protect intellectual property, eBay isn't going to be there. Call it the Globalized Club. No eBay for Indonesia, Nigeria, or Russia any time soon. Delivery problems may keep it out of India, banking scandals out of Malaysia. But eBay will not be dominant in *all* the nations listed: Yahoo controls 95 percent of the on-line auction business in Japan.

One striking and very American feature of these websites is their rating or feedback systems. This habit of "rating" service and goods, developed earlier by organizations like *Consumer Reports* and J.D. Powers, is spreading rapidly beyond eBay, Amazon, and other e-merchants. Not only do private enthusiasts' websites rate everything from cameras to Cameros, but generalized rating sites such as Epinion.com (owned by Shopping.com Ltd.) solicit feedback on practically any merchandise or service sold. The mere collation of opinion now equals clout.

Amazon.com was first to post user reviews of books prominently and then of other products. Because it had sales of $5.3 billion in 2003, Amazon's reviewers are potentially powerful, and their importance was underlined when Amazon in 2002 put in place reviews of reviewers, asking buyers how useful they found the opinions. Academic research suggested that negative reviews were more powerful than positive ones, which readers assumed to be publishers' puffs. In 2004 Amazon began author verification procedures to prevent hyping or slamming and to cut the number of anonymous reviews. Not only ratings, but this kind of cybernetic feedback with looping will no doubt spread.

EBay has a particularly complex rating system for its sellers and buyers, through which its marketplace is mostly self-policed. In most nations, its rating system works exactly the same as in the United States. But here again, academic research showed that negative feedback could be devastating, sending sellers into a downward spiral and out of business. To see the transformative power of reviews, check out eBay Mexico, where the negative ratings for sellers are quite high compared with those in the United States and the repartee between sellers and buyers is spirited, to say the least. In Mexico, Spain, and elsewhere, sellers often respond to positive feedback with praise of their own—"estoy para servirte" (I am here to serve you)—rather than the canned sentiments typical of eBay United States.

Comparison shopping sites similar to the American BizRate.com, PriceGrabber.com, and Froogle are also popping up abroad. Though they include consumer feedback, an equally potent feature called "tags" is unseen. To get their merchandise listed at comparison sites, commercial websites have to use "tags" that can be picked up by the web-crawling software of the comparison shopping sites. They have to update prices, shipping charges, and inventory daily. Without this "transparency," sites risk disappointing customers, who leave nasty feedback. E-retailers in the United States now pay between $5 and $10 per $100 of sales for comparison-site-generated sales, not to mention a few pennies per pass-along click from sites like Google. So there are financial pressures to conform to these protocols, to standardize product information and presentation, and to be honest and accurate.

And these features have their intended effect abroad. At the Chinese website of NewEgg.com, consumers praise and disparage products just as candidly as Americans do at U.S. sites. With a Chinese colleague, I looked at NewEgg reviews of a Chinese-made joystick for computer games. Opinions varied, with several customers complaining about the difficulty of setup, while others responded with helpful hints. Some said the joystick was just as good as imported models, but others said it proved that one should "never buy anything that says 'Made in China.'" These service or product ratings are similar to the American template in allowing Internet users to post anonymously, to enter written evaluations, and to assign from one to five yellow stars to the item.

At another site (Yongle.com) there was similar feedback on cell phones. The tone of the comments, translated for me, would be familiar to most Americans: alternately acerbic, technical, or bland. The most interesting comment came from a consumer who, having puzzled his way through the product setup and discovered its shortcomings, wrote: "I wish I had been able to talk to my big and little brothers here before I bought this item." By "big and little brothers," according to my informant, the consumer meant the traditional process of consulting within the extended family about the reputation of a product or merchant before making a purchase. Before buying a television, a Chinese consumer would ask within his or her family, beginning with older brothers, about what model to buy, from which merchant. The family might call more distant relatives or friends for recommendations, or contacts, or for merchants from whom some discount might be had. The consumer quoted here seeks to blend this older custom with the website feedback model.

Most academic attention to the Internet in China has focused on that nation's blockade of news and information. But Jonathan Zittrain and Benjamin Edelman, in an "Empirical Analysis of Internet Filtering in China," show that China's censorship is porous and spotty. Fewer than 10 percent of the 204,000 websites they tested were inaccessible from two or more points in China. The offending websites usually focused on Taiwan, sex, entertainment, health conditions in China (at the height of SARS), and religion (such as Falun Gong or Tibet) or were news outlets such as the BBC and *Time*. Always available were LexisNexis, Westlaw, Findlaw, *Hustler,* and whitehouse.com. Entrepreneurs are everywhere on the Chinese Internet: recent articles in the *Wall Street Journal* have highlighted the growth of both spam and blogs. Even Vice President Dick Cheney's speech in Shanghai, highly critical of China, was to be found entire if one played with the characters in his name. And Reporters sans Frontières comments that "the tremendous growth of the Internet [in China] now makes it technically impossible for the authorities to monitor the content of all the millions of e-mail messages being exchanged."[29] Another comprehensive comparison of the Internet in India and China by four scholars has concluded that in all areas of infrastructure and even "rule of law" and "trade policy," China was a more progressive Internet nation than India.[30]

Internet "blogs" in China address a range of contemporary topics quite freely. Militant nationalism, as well as anti-Japanese and anti-American rhetoric go unchecked at these sites. News and opinion dissemination also work through text messaging on cell phones, which is the fastest-growing area of business not only for Chinese cell phone companies but also for Internet portals such as Sina.com. The 2003 cover-up of the spread of SARS by the government was forced into the open by a crusading doctor who spread his opinion in this fashion. Later in 2003 a migrant worker was reportedly beaten to death in Guangzhou by police when he lacked proper identity papers. First reported by a local newspaper, and then suppressed, news of this case spread nationally through an Internet posting that was forwarded through millions of cell phones. The government was forced to change the law that permitted the worker's detention and led to his death.

Logistics

If space permitted, chapter 3 would also cover cleanliness, air travel, and tourism. For I have been struck, in humble Mexican *cantinas* and remote Jordanian tea shops, by customers who will line up at a sink on the wall to wash. In Indonesian *kampungs,* I have seen villagers walk a kilometer to bathe in the evening. People everywhere want to be clean, but the technology of bringing water to houses and carrying waste has usually accompanied the "American" part of modernity. The United States has also been the source of myriad products that make houses, restaurants, and workplaces cleaner. Procter & Gamble, as a recent study shows, had demonstrably superior products in "Tide" and "Ivory," to which the world responded.

But space does not permit such exploration, and the implication of my treatment is by now clear. As Richard Pells suggested in *Not Like Us* (1997), the real "American" face of globalization probably consists of "methods," what we now call *logistics.* The word dates to 1875, when it pertained to the military: "the procurement, supply, and maintenance of equipment." By the 1970s it had a second definition: "the planning, coordination, and implementation of the details of a business." It is a commonplace of American history that their frontier experience made Americans adept at transporting, communicating, and organizing. It is less often appreciated how dynamically new technology has been reinforced by this cultural legacy. Logistics is the part of globalization at which Americans still excel.

Logistics is a profession that is growing in the United States even as other kinds of jobs disappear. "Companies are hiring logistics consultants to untangle their supply chains and to monitor shipping lanes and weather patterns," writes Robert Guy Matthews in the *Wall Street Journal.* MIT has expanded its logistics program and started a new master's degree program in its School of Engineering. Jarrett Logistics Systems, in tiny Orrville, Ohio, tripled its revenues in 2004, after tripling them in 2003.[31]

Beyond the complexity of integrating the various systems outlined in this section, the demand for American logistics expertise is being driven by two forces. One is the worldwide demand for product variety. A product introduced from Austria like the energy drink Red Bull can make obsolete a product seven thousand miles away (Jolt Cola, for instance). A product with a 20 percent promotional discount at the store can increase demand by tenfold, but can the manufacturer produce and deliver enough? No company wants to tell vendors it ran out of the promotional item. Such complexity is actually driving the convergence of logistical systems. Even the financing of international orders, which used to be settled on paper with letters of credit (essentially unchanged since the day of the Medici), has gone "just-in-time." Banks used to take up to two weeks to clear letters of credit, adding as much as 5 percent to the cost of trade. Then UPS and other transporters decided they could easily carry payment back after making deliveries. U.S. shoe manufacturer Hi-Tec says this saves 20¢ on the $16.50 manufacturing cost of a pair of boots this way. Another force pushing logistics into prominence is international uncertainty. Terrorism, wars, earthquakes, floods—these events loom larger when the world is a global village. If oil stops flowing after Hurricane Katrina, if the garment industries of Afghanistan and Pakistan shut down, if riots close the port of Pusan, if natural gas no longer flows from Australia—how does everyone adjust?

The "ways of doing things" that Americans have pioneered are becoming the ways that millions of people elsewhere do things. From ATMs and container ports to airfreight and bar codes, technologies pioneered in the United States have been adopted by other nations, which are now going into competition with and sometimes surpassing the United States. In this I see a second major implication. As the case of the Indian ATMs shows, local adaptations of U.S. technology are often locally superior. This sets a limit to the advance of certain Americanisms, such as the drive-through ATM or burger joint. Other Anglo-American practices, such as using the money market, allow Indians to invest in money market funds, which reduces the advantage Americans have had in accumulating wealth. The upshot of this appears to be that some parts of the developing world will improve their lot dramatically. Sharing a common logistical skill set with us, nations like India and China will be drawn closer to the United States in some ways, as they take practices like franchising or just-in-time manufacturing but customize them for their use.

On the other hand, as John McPhee showed in a 2005 profile of UPS in the *New Yorker,* Americans remain far in the advance of logistical invention. The competition between FedEx and UPS (the seventh and eleventh largest airlines in the world by revenue) has spurred the latter to a level of creative systems integration that no overseas company even imagines. "Now they have Worldport, as they call it—a sorting facility that requires four million square feet of floor space and is under one roof. Its location is more than near the Louisville International Airport; it is between the airport's parallel runways on five hundred and fifty acres that are owned not by the county, state, or city but by UPS," wrote a plainly astonished McPhee.[32] He detailed how live lobsters from coastal Maine were *trucked* all the way to Kentucky to be *flown* all over the world, including back to Boston. He described the 1 million items per day sorted between 11 p.m. and 4 a.m.—bull semen, auto mufflers, and baseball bats among them—through the windowless, seventy-five-foot-high central "sort."

> This was the Grand Canyon of UPS. On each of ten or fifteen levels, packages were moving in four compass directions at the rate of one mile in two and a half minutes on a representative sampling of seventeen thousand high speed conveyor belts. Pucks were pushing packages to the left, to the right, including lobsters that raced into cylindrical spaces and whirled in semicircles as if they were on an invertigo ride with an "aggressive thrill factor," in the language of amusement parks. In no other place could you absorb in one gaze the vast and laminated space where, in the language of UPS, "automated sortation takes place."[33]

UPS has made it *more efficient* to truck the lobsters from Maine to Kentucky to send them to Boston, as well as faster and more dependable. Some of them will be on tables in Puerto Rico, Paris, and Tokyo the next day. But this extraordinary, Escher-like facility is not the whole story. At the same time that UPS centralizes transport for some items, it eliminates it for others. As McPhee explains,

> If you own a Toshiba laptop and something jams, crashes, or even goes mildly awry, you call 1-800-TOSHIBA and describe your problem. If the answerer can't help you, a brown package car shows up at your door with an empty padded box hollowed out in the shape of your laptop. UPS takes your computer overnight to Louisville, and keeps it there. Two miles south of the runways are six more UPS buildings, white and windowless in a spotless and silent landscape campus, and covering, on average, more than three hundred thousand square feet. Your laptop goes in there—Building 6. Within a few hours—in a temperature-controlled, humidity-controlled, electrostatic-sensitive area—an electronic repair technician who is a full-time UPS employee will have the

innards of your Toshiba laptop spread all over a table. . . . In a day or two, your laptop takes a ride through the sort and flies in a browntail back to you.[34]

UPS performs similar services for Rolls Royce, Nikon, Nokia, and Bentley. The Bentley factory in England has called this UPS facility to send it parts. Many Amazon.com and Jockey products are prepositioned in UPS warehouses. Hillerich & Bradley, which makes Louisville Slugger baseball bats, ships a hundred a day to assorted destinations. Its UPS bill is sometimes $30,000 a week.

At the other end of its operations UPS uses a device similar to the FedEx codabar. UPS calls it the delivery information acquisition device or DIAD ("die-add"). It's bigger than the codabar, but it can generate labels. McPhee tells of a sixty-nine-year-old man who decided to walk from New Jersey to California. He planned his trip in such detail that he had UPS bar-coded T-shirts, socks, and shorts awaiting at each stopover. He would meet a UPS driver when he arrived and be scanned. He was still walking when he turned 70 and UPS delivered, on the road, his birthday cake.

Logistical expertise is a key factor—maybe *the* key factor—in increasing productivity. It explains why, despite competition from low-wage countries, the productivity of American workers continues to lead the world or to be close to the top of the heap. As Dani Rodrik notes, there is a long-standing correlation between labor productivity and labor costs, with the United States, France, Germany, Japan, Britain, and a few other nations consistently ranking high in both. If it were true that low wages were driving globalization, he remarks, then Bangladesh and some African nations would be the world's most formidable exporters. "Some Mexican or Malaysian exporting plants may approach U.S. levels in productivity, while local wages fall far short," Rodrik concedes. "Yet what is true for a small number of plants does not extend to economies as a whole."[35] Productivity and wages rise in tandem, with logistical expertise as a key link.

Conclusion

> What if people abroad do not want to be just like us? What if they adopt
> our methods, buy our products, watch our movies and television shows,
> listen to our music, eat our fast food, and visit our theme parks, but refuse
> to embrace our way of life? What if they insist on remaining "foreign,"
> un-American, even anti-American?
>
> —Richard Pells, *Not Like Us*, xiv

The idea that the globe is being "Americanized" has been around for
more than a century. In 1902 William T. Stead, a reform-minded English
journalist, wrote a book titled *The Americanization of the World*. A thor-
oughgoing internationalist, Stead wanted Britons to overcome their
snobbishness and to join with their former colony to enlighten the world
culturally and financially. It would be a "race union" of English-speaking
peoples, seated in the fatherland. In his book Stead reviewed the advantages and disadvantages of New World ways to his scheme. He admired
Americans' reciprocity in trade—"benefits must be given if benefits are
sought"—as well as the Monroe Doctrine, freethinking Protestantism,
and the Americans' "host of ingenious inventions and admirably perfected machines which we are incapable of producing for ourselves."[1]
However, he saw rather peculiar summits in American culture, singling
out novelist Harriet Beecher Stowe, the painter Edwin Abbey, and marriage (noting that American wives were unusually common in the British
cabinet).

Nevertheless Stead was on to one thing. As Inglehart, Hofstede, and
others have since shown statistically, prosperity has spread to democratic, Protestant, English-speaking economies much more rapidly than to

the "Rhineland" and "Mediterranean" models, not to mention those of Asia. These cultures were more open to "modernity." But we shouldn't be misled into thinking that Stead was in any way prophetic. People had been using *Americanize* as a verb for 120 years before Stead suffixed it. Shortly after the Revolutionary War, it was used to describe two aspects of the new nation. On one hand, the united colonies needed to create a common culture and to embrace further self-fashioning. Even Benjamin Franklin underwent this process, as Gordon Wood details in *The Americanization of Benjamin Franklin* (2004). To *Americanize* was also an internal dynamic of new citizens, carried forward by discussion, debate, and simple expediency. They needed systems of government, roads, trade, schools, and social conventions. What would these be? The strongest sense they had of themselves as "Americans" came from the contrasts they felt to their homelands (if they remembered them) and to newer immigrants, who were a serious problem. The latter soon became the major focus: how could the foreign-born pouring into the country be naturalized? From 1790 onward, dictionaries tell us, to *Americanize* meant to acculturate.

A second meaning of *Americanize* developed abroad, for the world watched the new country curiously. *Americo-mania* was a term used in the British press from 1798 through 1880. Foreign observers arrived and reported home that Americans now had certain distinct "characteristics," which one might call a culture. Glimpsing themselves in foreign mirrors, Americans took up the topic, with Crèvecoeur and Jefferson in the lead. Then the visiting Tocqueville, Dickens, and Mrs. Trollope wrote books on the curious ways of Americans, stressing their practicality, unpretentiousness, or bad manners. By the mid-1800s, to *Americanize* was common parlance. Novelist William Dean Howells offered no explanation when he wrote in 1875 that one of his foreign characters "was Americanizing in that good lady's hands as fast as she could transform him."[2]

These two uses continued into the early 1900s. For U.S. citizens, *Americanize* meant to acculturate immigrants through English classes, cooking lessons, and instruction in hygiene, which were delivered by government, women's groups, and religious charities. They undertook such activities with evangelic zeal, contributing to cultural nativism and diplomatic isolationism rather than any colonial exploits. But for Stead—who never visited the United States—Americanization consisted of externally visible cultural and economic traits. The United States emphasized education, personal incentives, freethinking in religion, and democracy. Unlike most European nations it could feed itself, and it embraced labor-saving machines. Like China today, it was then flooding the world with "cheap goods" (which Stead courageously defended). By linking up with such a culture, he argued, Anglo-Protestantism could be extended not only to Ireland, South Africa, Canada, and Australia, but throughout the Americas, Asia, and Africa—while headquartered, of course, in Britain. There was no sense in *The Americanization of the World* that American culture, popular or otherwise, was exportable, and Stead made scant mention of the McCormick reaper, the transcontinental railroads, the telephone, the stock and commodity exchanges, the automobile, or other American technological advances.

World War I produced a new understanding of Americanization. European allies turned first to the United States for food and matériel, becoming aware not only of deep resources but also of American logistical prowess. Men followed, millions of them, a debt that more conscientious Europeans realized could never be repaid. As Luigi Barzini and Paul Fussell have pointed out, contact with U.S. soldiers changed forever the remoteness associated with America. Now it had a human face, which loved jazz and films and cars, which detested filth and pessimism. But the reparations and war debts that followed—administered by American banks and administrators—made those consumer goods hard for Europeans to obtain. As Richard Pells has argued, European intellectuals especially resented the ways in which their lives seemed diminished after the war.

It is not surprising, then, that academic studies of Americanization often begin in Europe during this period. But aside from a little jazz and film, Americans hadn't left much of their culture behind between 1917, when the United States entered the war, and 1932, when most of "The Lost Generation" departed. Studies that attempt to show otherwise usually end up choking on the exhaust of Teddy Roosevelt's speeches or vaporous "influences." But that period *was* the entering wedge of *modernity* in Europe, as the "masses" pressed for indoor toilets, stoves, autos, tractors, and phonographs. They had seen and heard that Americans had these things. In order to achieve comparable material progress, European governments and businesses began to blend in American models. As Ton von Schaik has written, "The economic successes of the US in the past century triggered imitation by other countries, not only imitation of American technology but also of the American rules of the game. Imitating, however, is not to be understood as copying literally. New rules melt gradually into existing rules, which are inherently linked with the cultural heritage of a country."[3] For some European intellectuals, it didn't matter whether the change was domestic "modernization" or "Americanization." The changes ran counter to Marx, who had predicted that the masses would arise to overthrow their masters. To such critics it seemed that governments had discovered new and unfair methods of preventing the millennium. They developed a special distaste for American popular culture, which they said created a "false consciousness" of well-being, and they grafted this on to their disappointment.

In the 1930s some of these critics, such as Max Horkheimer, Theodor Adorno, and Herbert Marcuse, emigrated to the United States to escape Nazism. The first two returned after the war and became central to the Frankfurt School, where Horkheimer attacked the "corruption of local traditions" by professional sports and the "loss of local autonomy" in mass communications. Adorno famously disparaged American music: "Jazz is a form of manneristic interpretation. As with fashions, what is important is show, not the thing itself. . . . Jazz, like everything else in the culture industry, gratifies desires only to frustrate them at the same time."[4] For the masses Adorno championed Schoenberg's twelve-tone scale. Marcuse stayed in the United States and taught at the University of California at San Diego, where he popularized the idea that this "culture industry" controlled things. Meanwhile Erich Fromm argued that advertising, not religion, was the opiate of the masses, stifling the critical capacities of the customer.

These critics, along with Louis Althusser in France and Antonio Gramsci in Italy, succeeded in emptying Americanization of its previous meanings and imposing their own stamp on it. Enter "Americanization" into Google.com's search engine today, and it will return more than 284,000 hits. Most of them are cultural "diagnoses" inflected by the Frankfurt School: the Americanization of aid to Africa, of British welfare, of German steel, of the Holocaust, of abstract expressionism, of British Columbia's forests, of pho, of reggae. Most people have accepted the neo-Marxian conflation of modernity, globalization, and Americanization. Lawrence Grossberg asks at *Keywords* (an update of Raymond Williams's Marxian lexicon) if globalization isn't "American capitalist culture replacing and destroying all local and indigenous forms of cultural expression?"[5]

This misperception arises from a particular set of cultural circumstances, which we need to get beyond in order to see globalization for what it is and what it is not. Curiously, Americans are unaware of globalization at home. But Americans traveling abroad, especially critics, tend to see Americanization everywhere. Yet by definition globalization is occurring globally. For most of us globalization is never about the changes occurring to *us*, but about change in Others. Theorist Wolfgang Iser has written that "in a rapidly shrinking world, many different cultures have come into close contact with one another, calling for a mutual understanding in terms not only of one's own culture but also of those encountered. The more alien the latter, the more inevitable is some form of translation, as the specific nature of the culture one is exposed to can be grasped only when projected onto what is familiar."[6]

We can see this "translation" in the writing that conflates modernity and Americanization from Stead onward. Stead used America to project change onto the late British Empire, and critics today use Ghana or Thailand to project change onto the United States. In the "marketplace of interpretation," as Iser calls it, these critics work out of their own preexisting personal *habitus* in order to translate their own relative advantages into something like penitence for like-minded readers. It has become a genre—the dystopia of modernity—in which Western "materialism" leads to personal lack of fulfillment *because* masses of "subaltern" peoples work in "sweatshop conditions" to make shoes or chairs or computers.

That's most of what we see when we Google "Americanization" plus anything: some version of the dystopia of success. Unfortunately, this genre provides a narrow lens through which to see the reality of broadband, instant *ramen*, and container ports. The result is that foreign cultures are not analyzed in their fullness, but reduced to a narrative function of this genre. We don't consider how Thais perceive logos, their habits of eating, of land use, of education, of saving and spending; their development and use of food technology, marketing, or their attitude toward the foreign. We have overlooked the *reception* of globalization.

Everywhere in such writing we find the assumption that other cultures are changing too rapidly for their own good. There is a kind of sympathetic projection of the critic's self onto the foreign culture. As Iser notes, "If the experience of crises, issuing into a critique of one's own culture, is meant to balance out the deficiencies diagnosed, recourse to other cultures proves to be a means of therapy for a growing awareness of cultural pathology."[7] This therapy locates self-reflection in the "stricken" foreign culture, which can then be used to redress oneself critically. There is a striking example in an academic book on Thai prostitution, in which the (male) author writes that "the [johns] look pink and flabby, and in the tropical heat (90 degrees in the 'cool' season), I've become very sensitive to the strong *farang* body odor, wondering if I smell like that too."[8] This study not only blames prostitution on globalization but also links it to Western concepts of masculinity.

In this case and others, the cultural pathology is not globalization, but the domestic dis-ease of educated Westerners with the speed of change. The cultural capital of scholars and critics lies in knowing traditions. In their new travels, the rest of the world no longer looks traditional or Other enough. Wherever they go, there are recognizable *signs:* Visa, Toyota, and Adidas. Mickey Mouse grins at the visitor in France and Japan. Taxi drivers and the hotel employees speak a little English. This is not what the educated Westerner goes abroad for, which is to get out of oneself and to encounter the Other. Hence the critiques tinged by a tone of regret for a past and an Other presumed to be fading. "Don't change," it implores, "Remain as you were." Implicit here is the critic's nostalgia for a lost personal authenticity. But what the Other's culture *was,* in all its premodern unpleasantness, and the critic's motive in reifying it, are never examined.

Earlier, in discussing the assumptions at work in such writing, I cited John Tomlinson's analysis. He reminds us that underlying the spread of globalization "is the spread of the culture of modernity itself" and that "the 'imaginary' discourse of cultural identity only arises within the context of modernity."[9] Perhaps this is too kind to those critics who, in his example, want to deny the Bushman a television. It is not the Bushman who has changed so much but the critics. This instantiation of change in the Bushman is best understood in the context of the critic's environment. In only 1980 few of them used computers, much less e-mail. The Xerox was expensive, there was no foreign travel budget, no DVD, no 401(k) retirement plan, no frequent flyer miles, and no Amazon.com. No one had heard of a leased auto or a "significant other." In fact, *no lives have changed as much in the past twenty-five years as those of Western critics of globalization.* What can it mean to have accepted such changes oneself, to most appearances uncritically, and yet to denounce their appearance in Others? Can someone located in a culture dominated by swift change even perceive accurately what is changing in a foreign culture?

The rate of change in the critics' lives has led to a sort of constant and recursive monitoring of the known as it appears in the Other. Is it a desire to slow the rate of change, to find a more stable subjectivity for oneself, that leads, paradoxically, to sightings of oneself and one's culture everywhere? Isn't this critique really a nostalgia for *authenticity,* which is always located outside the self? The critic implies the dire consequences of watching television, using computers, and eating at KFC. Of course, the critic could give up these things, but it is somehow more edifying to construct an Other who remains untainted. We never read that the Other should resist toilets or vaccines but rather the temptations of Disneyland and Citibank. Who is tempted here?

To understand how globalized the United States has become, let's look at Ohio. This "Rust Belt" region lost tens of thousands of jobs after 1980, as steel mills and auto plants closed. Today about a thousand foreign companies have subsidiaries there, employing 200,000 local residents. Sixteen thousand Ohioans build most of the Accords driven in the United States at a $5 billion Honda facility. In 2004 the Japanese firm spent more than $7 billion buying auto parts from 175 suppliers in Ohio. Not only has it not laid off workers, it has hired many new ones. The average hourly rate of its nonunion workers is $23.20, excluding overtime and bonuses. Studies show that foreign companies pay 19 percent higher wages on average than American companies do. Throughout Ohio, foreign companies hired during a period (2000–2004) in which the state lost 225,000 jobs. The jobs were lost due to a variety of causes, mostly technology, but they were not replaced by American companies as much as by foreign companies. When General Motors spun off its parts-making Delphi subsidiary near Columbus, forcing it to compete in the world market for business, employment dropped from 5,500 in the 1970s to 860 in 2004. But Honda and new suppliers picked up the employment slack. In this respect Ohio represents the domestic face of globalization. Nationwide, foreign companies have made direct investments of $1.31 trillion in U.S. operations since 1994, while U.S. companies have made direct investments of $1.19 trillion in foreign operations, a net influx of $120 billion to the United States.

This globalization was key to the U.S. economic recovery after 9/11, and knowledge about how to accommodate it, I would argue, constitutes a new form of American cultural capital. For the most part, foreigners have purchased *existing* businesses in their areas of expertise and then expanded them. In northwest Ohio the leading gasoline chain is now British Petroleum, which bought Sohio, and chief among its rivals is Shell (owned by Royal Dutch Shell). The dominant grocery chain is Tops, a unit of Netherlands' Royal Ahold, which faces low-end competition from Germany's Aldi stores. Another Dutch firm, Reed Elsevier, bought LexisNexis, the Ohio database company, in 1994. The Luxottica Group of Italy bought the local eyewear chain LensCrafters in 1995. Comp USA, owned by a Mexican family, is now Ohio's leading computer vendor. Cement for highway and building projects comes from Cemex, a Mexican company. Henkel of Germany bought adhesives maker Manco in 1998 and OSI

Sealants in 2004. The Royal Bank of Scotland bought Charter One Financial, a major Ohio bank chain, for $10.5 billion. The 7-Eleven stores in Ohio are, of course, controlled by Japan's Ito-Yokado. The Circle K's, Dairy Marts, and Dunkin' Donuts in northern Ohio are owned by Alimentation Couche-Tard of Canada. The Crate and Barrel stores and Spiegel are owned by the Otto Group of Germany. The Sbarros and Burger Kings are owned and operated by Diageo, a British company. Cleveland's steel mills had Japanese, then Brazilian, and now German owners. The venerable Ingersoll Milling Machine Co. was purchased by Camozzi SpA of Italy in 2004, which fortunately kept it in Ohio; not only is the tool-and-die business essential to national defense, but it employs thousands of Ohioans. On my university campus, the faculty order books through Random House (Germany's Bertelsmann) and the food services are provided by Sodexho (France). Global companies have brought new expertise and new ways of doing business to Ohio.

The United States is already more globalized than any other nation. But even its educated citizens don't realize the extent to which they themselves have *chosen* to globalize. Take local grocery stores, where we might suppose that our choices were predominately domestic. We think we recognize the "foreign." That Dannon yogurt is French, and those Goya products are Mexican, right? Actually, all Dannon yogurt sold in the United States is produced at three U.S. plants, including one at Minster, Ohio, that turns out 3 million yogurt cups a day. Dannon does $18 billion a year in business in the United States. When we reach for a Lipton, Hellmann's, Birds Eye, or Slim-Fast product, do we realize that we're buying from Great Britain's Unilever? It makes more than a thousand brands of food that are sold through three hundred subsidiary companies in eighty-eight countries worldwide—brands such Knorr, Ragu, Bertolli, Dove, Pond's, Lux, and Surf that we use every day.

Most of us don't realize that we're shopping at foreign-owned stores or that we're buying foreign-produced products. Recently at my grocery, I bought a bag of Archway Ginger Snaps. Great taste and very crisp, reminding me of Italian *biscotti*. And no wonder. When I read the label, I learned they are baked in California by Parmalat, an Italian company, which makes the Archway, Salerno, Delicious, Frookie, and Mother's brands of cookies. Parmalat also produces the Beatrice, Lactaid, Soy, and Parkay product lines.

If I were properly suspicious of globalization, I would also stay away from Dr Pepper and 7 UP—those are British (Cadbury-Schweppes). I'd avoid Miller beer (SABMiller of South Africa) as well as Labatt's, Rolling

Rock, and Corona (Belgium's Interbrew). And no Lifesavers, Lean Cuisine, Alpo, Friskies, Hill Brothers, MJB, Dreyer's ice cream, or Stouffer's products (a few of the thousands of Nestlé brands). In the magazine aisle, I'd avoid *Car and Driver* (LaGardère, of France) and *Hot Rod, Guns & Ammo, Snowboarder, Motor Trend,* or *Teen* (E-Map of Britain) as well as *Family Circle, Parents,* and *Child* (Bertelsmann of Germany).

Just for contrast, let's return to that French Géant Casino grocery discussed in chapter 2. Remember how the "foreign" foods were rigorously separated in separate aisles? In the "United States" aisle were Skippy peanut butter, Kellogg's corn flakes, Uncle Ben's rice and pancake mix, Heinz ketchup, A-1 steak sauce, and Gallo wines. There was a sprinkling of Lever products from Britain, which we might be misled into thinking American, and a dozen canned goods from German firms. The world's second largest economy, Japan, was represented by Kikkoman soy sauce and Sapporo beer. This sector of the French economy is rigorously protected from globalization. Has it retained more local culture? Possibly. Are the French better off? They spend twice as much of their disposable income as Ohioans do just feeding themselves. Their unemployment rate hovers around 12 percent. No one is rushing to invest in the French economy.

Back in the Rust Belt, Ohioans have gained the cultural capital of globalization, which is a close knowledge of how and when its bilevel system works for and against them. For example, there are unionized American and nonunionized foreign auto plants in Ohio, and there is a labor force that is conscious of the pluses and minuses of each workplace, even if they haven't worked there. There are thousands of designers, tool-and-die workers, and sales people who work for large and small auto parts suppliers that compete for American and foreign auto companies' parts contracts. The intensity of this competition and the efficiency demanded of them in order to reduce costs is stunning to witness, and it has created an ability to design quickly, adapt quickly, and to eliminate steps in processes. Thomas Friedman has called this a "flat world," but that's wrong. The metaphor is anachronistic. There's nothing flat about a world in which 60 percent of China is not connected to international transportation and the United States ranks fourteenth in Internet broadband connections. Rather we are approaching a bilevel world—multilocal and confluently global at the same time. In places like Ohio, the increased foreign presence internationalized the consciousness of Ohioans, who now understand differing manufacturing processes, standards, and sizes, different product uses and packages. The foreigners have shown Ohioans what they do best, and synced them up with global buyers and sellers. They have also augmented Ohioans' cultural knowledge: sushi is Japanese, dim-sum Chinese, pho is Vietnamese. Everything we ship UPS goes through Louisville, which is not so far away. This is the database template they use in Germany, and Hondas made in Ohio can be just as good as Hondas made in Japan.

Having been globalized in this fashion, Ohioans increasingly understand what they do well by international standards, what they raise or produce that is unique, and what they have to do to stay at that level. They no longer measure themselves by local standards, the clearest indication that, impossible as it would be for most of them to express it, a new form of cultural capital is forming. A striking example was the announcement in August 2004 of a new "in-sourced" auto plant in Toledo. Three auto parts suppliers—two of them foreign—will collaborate with Daimler-Chrysler to build a $900 million Jeep plant. The suppliers will own the plant, Hyundai will supply the chassis, and Chrysler will build the Jeeps. "The concept has been tried overseas," said officials, "but it's new to the United States."[10] And it can't be done just anywhere; it requires cultural capital. Albeit intuitively, Ohioans understand direct foreign investment—it works!

The primary concern of critics of globalization has been that national cultures will become "standardized." Differences, they say, will be "leveled out" and products "homogenized." But it is difficult to think of egregious examples here in the United States. In fact, our own experience suggests that there will be more, rather than fewer, products and cultural choices. As I tried to show in chapter 3, it is the logistical systems that deliver them—financial services, container shipping, airfreight, computing—that will move toward standards, such as technological efficiency and economies of scale. These enable work *across* cultures. To the extent that logistical systems can, they accommodate local culture. For example, the language interfaces of ATMs can be local, but their logistical systems will be global. The cultural "local" endures.

There is often a misleading appearance of standardization where there is none. As discussed in chapter 2, most products or practices must be adapted to the tastes of the local market, which has deeply rooted ways. Thus we get the American version of "ramen," and the Japanese get their version of "cake." Even McDonald's "quarter-pounder," as Quentin Tarantino reminded us in *Pulp Fiction,* is a "Royale with cheese" and tastes differently in France. In Indonesia McD mostly sells very spicy chicken. It is not fast-food sellers who threaten to standardize cuisine, since their

trade depends entirely on cultivating local customers. The standardization is in the production and delivery systems of conglomerates such as
ConAgra, Unilever, Nestlé, and Royal Ahold.

Yet even there, consumers are continually reinterpreting products.
After a foreign product or practice has been domesticated, it can even be
reintroduced to its market of origin in this new form and catch on again.
A case in point is tequila, a drink of humble origins in Mexico. Middle-
and upper-class Mexicans wouldn't touch the stuff in the 1960s. But
extranjeros took an interest and began to produce it abroad, increasing
the quality of the ingredients, refining the taste, and raising the price. In
the 1980s Mexicans traveling in Europe discovered the designer tequilas
and demanded them at home, creating a new market for a traditional
drink.

Very little of this product diversification is culturally destabilizing.
Even the logistical system with the most impact—the money market—
has not shown itself capable of altering the foundations of folkways. The
sudden outflow of funds from Mexico or Argentina, following their
crises, didn't change the diet or language or TV-viewing habits of those
nations' citizens. Argentines haven't given up meat for soy products, nor
Mexicans corn for potatoes. Instead the globalization that proceeds from
logistical technologies augurs a future that is, for lack of a better phrase,
smooth globalization. The production of cars and computers, the functioning of the Internet, the distribution of films and other media, the
flow of capital, and international travel and trade will function with
fewer bumps, devaluations, or delays. The gradual decline of international trade barriers, the rise of outsourcing, and new production in
lower-wage nations will lessen global inequalities in wealth and reduce
some of the pressures of immigration.

Smooth globalization seems compatible with local cultural particularization. Its technologies support local languages, music, foods, and film.
These are new market niches for its processes. Economists tell us that
handicrafts and service specialties, from *chocolatier* to *masseur,* will increase in the future. The United States is already witnessing this trend.
The seamless processes of smooth globalization will also create a demand
for their phenomenological opposite—the local, particular, or *rough* and
original culture. This desire for contrasts seems to be built into human
nature. Effortless ATMs allow Ohioans to buy rhubarb and basil at local
farmers' markets and Egyptian bankers to shop in the souk on the way
home. A day spent in front of a computer may make the Japanese worker
inclined to *koto* lessons in the evening. German engineers using CAD-

CAM all day take up pottery, woodworking, and watercolors in the evening. So far globalization has only smoothed the way to greater ethnic and language and religious particularization—in short, *greater diversity*. It has depended on reception and choice; it has promoted adaptation instead of imposition. How bad can that be?

Notes

Chapter 1. *"Less Than We Think"*

1. Jesse Katz, "Garment Smugglers Turn U.S. Discards into Mexican Treasure," *Washington Post,* December 12, 1996.
2. Center for Strategic and International Studies, www.globalization101.org/issue/culture/28.asp.
3. Burton Bollac, *Chronicle of Higher Education,* September 8, 2000, A73.
4. Fredric Jameson, "Notes on Globalization as a Philosophical Issue," in Jameson and Masao Miyoshi, eds., *The Cultures of Globalization* (Durham: Duke University Press, 1998), 59.
5. Walter Mignolo, "Globalization, Civilization Processes, and the Relocation of Languages and Cultures," in ibid., 41.
6. Samuel P. Huntington, *The Clash of Civilizations and the Remaking of World Order* (New York: Touchstone, 1996), 62.
7. Ibid., 60.
8. Joshua A. Fishman, "The New Linguistic Order," in Patrick O'Meara, Howard D. Mehlinger, and Matthew Krain, eds., *Globalization and the Challenges of a New Century* (Bloomington: Indiana University Press, 2000), 439.
9. Ibid., 438.
10. Ibid., 439.
11. Peter Krouse, "Talking the Talk," *Plain Dealer* (Cleveland), July 7, 2003.
12. Ibid.
13. Ibid.
14. Lynn Hirschberg, "Backstory," *New York Times Magazine,* November 14, 2004.
15. David Puttnam, *Movies and Money* (New York: Knopf, 1998), 36–37, 29.
16. Ibid., 201–2, 239.
17. John W. Cones, *The Feature Film Distribution Deal* (Carbondale: Southern Illinois University Press, 1998), 55.
18. Marie Brenner, "Dino De Laurentis Conquers America," *New York Magazine,* October 21, 1992, 51–55.

19. Frederick Wasser, *Is Hollywood America? The Transnationalization of the Film Industry* (Urbana: Institute for Communications Research, 1994), 19–20.

20. Data from Nielsen EDI, reported in *New York Times,* November 4, 2002. Sony and Buena Vista have traded the lead for several years, with the latter achieving a 20% market share in late 2003.

21. Kristin Thompson, *Exporting Entertainment: America in the World Film Market, 1907–1934* (London: British Film Institute, 1986), 81.

22. Ibid., 140.

23. "Gente," *Reforma,* May 20, 1995, 4D.

24. Riera, in "Culture Watch," *News* (Mexico City), May 19, 1995.

25. "Culture Watch," *News* (Mexico City), May 17, 1995, 12.

26. Chang-Ran Kim, "Hopeful Multiplexes Spin Shaky Web, Bank on Hits," *Japan Times,* May 24, 2002.

27. Hirschberg, "Backstory," 91.

28. Alessandra Stanley, "Italians Can't Believe Ears: Movies Lose Their Voices," *International Herald Tribune,* September 17, 1998.

29. Puttnam, *Movies and Money,* 261.

30. Ibid., 274.

31. Ibid., 227.

32. "French Film Industry Fears Its Worst-Case Scenario," *International Herald Tribune,* December 27, 2001.

33. "Culture Wars," *Economist.* September 12, 1998.

34. Jameson, "Notes on Globalization," 58, 60.

35. "Culture Wars," *Economist,* September 9, 1998. See also O'Meara et al., *Globalization,* 454–78.

36. The Khmer stations have a website: www.tv9.com.kh/marketing_research.html.

37. The titles of some shows are in English, but the programming was all Japanese. Titles shown here in *romanji.* The top three shows of the families of my students at Kobe College in 2001 were as follows. Variety: (1) *Waratte iitomo,* (2) *Koimi Karasawagi,* (3) *Vo no moto.* Music: (1) *Music Station,* (2) *Utaban,* (3) *Love Love Aishiteru.* Cooking: (1) *Dochi no ryori show,* (2) *Chyu bo desuyo,* (3) *3 Minutes Cooking.* Drama: (1) *Tenkyoho no koibito,* (2) *Eien no Ko,* (3) *Nurse no oshigoto.* News: (1) *News Station,* (2) *Super Morning,* (3) *Tokudane.* Travel: (1) *Sekai Ururun Taizaiki,* (2) *Ainori,* (3) *Sekaimo Syasoukara.* Movies: (1) *Friday Road Show,* (2) *Saturday Wide Gekigyo,* (3) *Sunday Youga Gekigyo.*

38. Eric Pfanner, "Europe's Dashed Cable Hopes," *International Herald Tribune,* March 16, 2002.

39. Puttnam, *Movies and Money,* 282.

40. Michael Wines, "Moscow Journal: Wired Radio Offers Fraying Hope," *New York Times,* October 18, 2001.

41. David Nye, during question and answer session, European American Studies Association, Warsaw, Poland, March 21–25, 1997.

42. Divina Frau-Meigs, "The Cultural Impact of American Television Fiction in Europe: Transfer of Imaginary Worlds or Cultural Compatibility?" paper presented to the European American Studies Association, Warsaw, Poland, March 21–24, 1996, 5.

43. Ibid., 7, 8.
44. Ariel Dorfman and Armand Mattelart, *How to Read Donald Duck,* trans. and introd. David Kunzle (New York: International General, 1995; originally published 1984), 98.
45. John Tomlinson, *Cultural Imperialism* (Baltimore: Johns Hopkins University Press, 1991), 49.
46. M. Barker, in ibid., 43.
47. Michael Switow, "Philippines Is Cartoon Capital," *Plain Dealer* (Cleveland), October 10, 1996, A17.
48. Ibid.
49. Kunzle, "Introduction," in Dorfman and Mattelart, *Donald Duck,* 14.
50. George Ritzer and Elizabeth L. Malone, "Globalization Theory: Lessons from the Exportation of McDonaldization and the New Means of Consumption," *American Studies* 41, nos. 2–3 (2000): 97–118.
51. Ibid., 105–6.
52. Emiko Ohnuki-Tierney, "McDonald's in Japan: Changing Manners and Etiquette," in James L. Watson, ed., *Golden Arches East: McDonald's in East Asia* (Stanford: Stanford University Press, 1997), 173.
53. Ritzer and Malone, "Globalization Theory," 108.
54. Ibid., 107–8.
55. Frank Thomas, *New York Times Magazine,* December 31, 2000, B4.
56. Lee Hockstader, "Attack on the Big Mac," *Washington Post,* August 8, 1995, in Ritzer and Malone, "Globalization Theory," 107.
57. Erin E. Arvedlund, "McDonald's Commands a Real Estate Empire in Russia," *New York Times,* March 17, 2005.
58. Mooradian and Turow quoted in Amy Harmon, "Web Clickers Stick to the Tried and True," *International Herald Tribune,* August 27, 2001.
59. Wayne Arnold, "Malaysia's Internet Road Show," *New York Times,* August 23, 2001.
60. J. D. Biersdorfer, "Taking the Net Where a Phone Is a Luxury," *New York Times,* August 23, 2001.
61. John Varoli, "Russia Tries to Catch Up," *New York Times,* July 7, 2001.
62. Jameson, "Notes on Globalization," 64.
63. http://en.wikipedia.org/wiki/Transnational_corporation.
64. Jameson, "Notes on Globalization," 67.
65. Keither Damsell, " More Foreign Firms Seeking Czech Employees," *Prague Post,* April 6–12, 1994.
66. Brunswick, in Carol Hymowitz, "Foreign-Born CEOs Are Increasing in U.S., Rarer Overseas," *Wall Street Journal,* May 25, 2004.
67. Robert J. Antonio and Alessandro Bonnano, "A New Global Capitalism," *American Studies* 41, nos. 2–3 (2000): 57. They go on to state: "Their parent firms are almost always embedded in a national and institutional home base, where their socio-political connections and knowledge of local business, political, and cultural environments provide them with competitive advantages."

Chapter 2. The Resistance of the Local

1. Fredric Jameson, "Notes on Globalization as a Philosophical Issue," in Jameson and Masao Miyoshi, eds., *The Cultures of Globalization* (Durham: Duke University Press, 1998), 63.
2. Sherif Hetata, "Dollarization, Fragmentation, and God," in ibid., 282.
3. Mayumi Negishi, "Prada Japan's Italian Chief Knows the Merits of Vagueness," *Japan Times,* January 13, 2004.
4. John Tierney, "Baffled Occupiers, or the Missed Understandings," *New York Times,* October 22, 2003.
5. Pierre Bourdieu, *Distinction: A Social Critique of the Judgement of Taste* (Cambridge, Mass.: Harvard University Press, 1984), 79.
6. Ruth Benedict, *The Chrysanthemum and the Sword* (Boston: Houghton Mifflin, 1946), 82–83.
7. Sidney Mintz, *Tasting Food, Tasting Freedom: Excursions into Eating, Culture, and the Past* (Boston: Beacon, 1996), 21.
8. Magnus Pyke, "The Influence of American Foods and Food Technology in Europe," in C. W. E. Bigsby, ed., *Superculture: American Popular Culture and Europe* (Bowling Green: Bowling Green University Popular Press, 1975), 84–85.
9. "Gender Identity," *Encyclopaedia Britannica Micropaedia,* vol. 5 (1998), 172.
10. Geert Hofstede, *Cultures and Organizations: Software of the Mind* (New York: McGraw-Hill, 1997), 80–81.
11. Aihwa Ong, "The Gender and Labor Politics of Modernity," *Annual Review of Anthropology* 20 (1991), reprinted in O'Meara et al., *Globalization,* 253–77.
12. Celia W. Dugger, "Modestly, India Goes in for a Swim," *New York Times,* March 5, 2000.
13. Frantz Fanon, "Algeria Unveiled," in David A. Bailey and Gilane Tawadros, eds., *Veil: Veiling, Representation, and Contemporary Art* (Cambridge: MIT Press, 2003), 72–87.
14. Eimi Yamada, *Beddotaimu Aizu* (Tokyo: Kawade Shobou Shinsha, 1987), 13.
15. Agence France-Presse, "Analysts Blame Sephora's Strategy for Japan Failure," *International Herald Tribune,* November 30, 2001.
16. Suzy Meknes, "A No-Nonsense, No-Logo Approach," *International Herald Tribune,* September 11, 2001.
17. David Lehney, "A Political Economy of Asian Sex Tourism," *Annals of Tourism Research* 22 (1995): 373.
18. John Burdett, "Author's Note," in *Bangkok 8* (New York: Knopf, 2003), introduction.
19. Ronald Inglehart, *Modernization and Postmodernization: Cultural, Economic, and Political Change in 43 Societies* (Princeton: Princeton University Press, 1997), 327, 380.
20. Ibid.
21. Joseph J. Tobin, "Dealing with a Difficult Child," in Merry L. White and Sylvan Barnett, eds., *Comparing Cultures: Readings on Contemporary Japan for American Writers* (Boston: St. Martin's Press, 1995), 89–94.
22. Inglehart, *Modernization and Postmodernization,* 78.
23. Ibid.
24. Richard Lewontin, "Organism and Environment," in Harold Plotkin, ed., *Learning, Development and Culture* (New York: Wiley, 1982), 178.

25. Chris Oakes, "Programmers Share a Code the World Over," *International Herald Tribune*, March 15, 2002.

26. James Brooke, "Japan Farms: An Old Man's Game," *New York Times*, November 7, 2003.

27. Ibid.

28. Ibid.

29. Fallows, www.theatlantic.com/unbound/fallows/jf2001-06-21/.

30. David Yetman, "*Ejidos*, Land Sales, and Free Trade in Northwest Mexico," *American Studies* 41, nos. 2–3 (2000): 214.

31. Ibid., 216.

32. Ibid., 226.

33. *Economist*, November 30, 2002.

34. Grant McCracken, *Culture and Consumption* (Bloomington: Indiana University Press, 1991), 31.

35. In Howard W. French, "A Hard Time Leaving His Mark," *International Herald Tribune*, September 15, 2001.

36. Ibid.

37. President Wade, in Joseph R. Gregory, "In Mali, a Fragile Democracy," *International Herald Tribune*, April 26, 2002.

38. Kofi Annan, "The Politics of Globalization," address presented at Harvard University, Cambridge, Mass., September 17, 1998. Reprinted in O'Meara et al., 125–30.

39. Somini Sengupta, "Ethnic Dispute Stills Nigeria's Mighty Oil Wells," *New York Times*, April 1, 2003.

40. Annan, "The Politics of Globalization."

41. Maureen Meehan, "Israeli Textbooks and Children's Literature," *Washington Report on Middle East Affairs* 9, no. 99 (September 1999): 19–20, www.washingtonreport.org/backissues/0999/9909019.html.

42. Hugh Pope, "A Saudi Leadership Adrift," *Wall Street Journal*, June 30, 2004.

43. Marlise Simons, "Resenting African Workers, Spaniards Attack," *New York Times*, February 2, 2000.

44. Katharine Boo, "The Best Job in Town," *New Yorker*, July 5, 2004.

45. Subramani, "The End of Free States: On Transnationalization of Culture," in Jameson and Miyoshi, *Cultures of Globalization*, 149–50.

46. Larry Rohter, "Trapped Like Slaves on Brazilian Ranches," *International Herald Tribune*, March 26, 2002.

47. Thomas Friedman, *The Lexus and the Olive Tree* (New York: Anchor, 2000), 146.

48. Friedman, in *Lexus*, 156, interviewed Elbirt and describes the incidents.

49. World Bank report quoted by Seth Mydans, "Cambodia's New King Dances into a Land of the Absurd," *New York Times*, October 23, 2004.

50. Tim Weiner, "Corruption and Waste Bleed Mexico's Oil Lifeline," *New York Times*, January 21, 2003.

51. Associated Press, "Power Thieves . . . ," *Japan Times*, July 1, 2002.

52. Hofstede, *Cultures and Organizations*, 49–78.

53. Tim Phillips, "A Web of Corporate Cops and Robbers," *International Herald Tribune*, March 28, 2003.

54. Ovsyannikov quoted by Guy Chazan, "In Russia, Politicians Protect Movie and Music Pirates," *Wall Street Journal*, May 12, 2005.

55. Associated Press, "Hong Kong Rife with Piracy, Global Music Bosses Allege," *Plain Dealer* (Cleveland), November 5, 1999.

56. *New York Times*, June 10, 2003.

57. Sebastian Moffett, "As Japan Ages, Concerns Shift to Retirees, Debt," *Wall Street Journal*, July 16, 2004.

58. Daniel Altman, "Tax Evasion Lies at Heart of Argentina's Economic Plight," *International Herald Tribune*, January 2, 2002 (reprinted from *New York Times*).

59. Geoffrey Fowler, "The Advertising Report: China. Questions for Austin Lally," *Wall Street Journal*, January 21, 2004.

Chapter 3. *"More Than We Know"*

1. Jathon Sapsford, "As Cash Fades, America Becomes a Plastic Nation," *Wall Street Journal*, July 23, 2004.

2. http://inventors.about.com/library/inventors/blatm.htm.

3. Don Wetzel, quoted in ibid.

4. Ye Rong, quoted by Craig Karmin, "China Prods Its Consumers to Use Plastic," *Wall Street Journal*, December 4, 2003.

5. Eric Bellman, "In India, Small Savers Mean Big Profit," *Wall Street Journal*, July 7, 2004.

6. Walter Bagehot, *Lombard Street: A Description of the Money Market* (ch. 3, p.1), www .econlib.org/library/Bagehot/bagLom6.html. Bagehot wrote that there had been earlier banking in Italy and Germany, but these banks, he noted, made loans to governments and for wars, and to guarantee "good coin," which was often lacking. Of the "Englishness" of commercial banking he wrote: "Many things which seem simple and which work well when firmly established, are very hard to establish among new people, and not very easy to explain to them. Deposit banking is of this sort."

7. Friedman, *Lexus*, 115–16.

8. Saskia Sassen, "Global Financial Centres," *Foreign Affairs* 78, no. 1 (1999), quoted in Friedman, *Lexus*, 116.

9. Van Harte, quoted by Ian McDonald, "Funds Adjust to Volatile Markets," *Wall Street Journal*, February 2, 2004.

10. Friedman, *Lexus*, 127, 129.

11. Juliet Sampson, quoted in "Lessons Learned: Markets Stay Calm Despite Argentina Unrest," *International Herald Tribune*, December 2, 2001.

12. Cavallo, quoted in Clifford Krauss, "The Downfall of an Argentine Star," *International Herald Tribune*, December 21, 2001.

13. *Economist*, December 25, 1997, cited by Friedman, *Lexus*, 121.

14. www.franinfo.com/whatis.html.

15. Robert Rosenberg and Madelon Bedell, *Profits from Franchising* (New York: McGraw-Hill, 1969), 171.

16. Pui-Wing Tam, "Fill'er Up, with Color," *Wall Street Journal*, August 3, 2004.

17. www.chass.utoronto.ca/~bhall/hps282f/Lecture%2016%20Civil%20Aviation.htm.

18. "Overstretched Kmart Files for Bankruptcy," *South China Morning Post*, January 23, 2002.

19. Toby B. Gooley, "Containers Rule," *Logistics Management* 36, no. 2 (February 1997): 71.

20. Vince Grey, www.iso.org/iso/en/aboutiso/introduction/fifty/pdf/settingen.pdf., p. 9.

21. www.maerskline.com/vessels.

22. Daniel Machalaba, "U.S. Ports Hit a Storm," *Wall Street Journal*, March 10, 2004.

23. In Christopher M. Singer, "Neighborhood vs. Business," *Detroit News*, January 9, 2001, www.detnews.com/2001/detroit/0109/19/s06-296861.htm.

24. Tony Seideman, "Bar Codes Sweep the World," *American Heritage of Invention and Technology* 8, no. 4 (Spring 1993): 58. Also at www.inventionandtechnology.com/xml/1993/4/it_1993_4_toc.xml.

25. Collins, in ibid.

26. Roger C. Palmer, *The Bar Code Book: Reading, Printing, and Specification of Bar Code Symbols* (Peterborough, N.H.: Helmers, 1995), 12.

27. Juliana Liu, "China's Web Search Engines Set to Take on Google" (2004), www.forbes.com/technology,newswire, 2004/03/10/rtr129281, reprinted from Reuters.

28. Margaret C. Whitman, CEO of eBay, quoted by Nick Wingfield, "Auctioneer to the World," *Wall Street Journal*, August 5, 2004.

29. Reporters sans Frontières, "China," www.rsf.org/article.php3?id_article=7237.

30. Larry Press, William Foster, Peter Wolcott, and William McHenry, "The Internet in India and China," *Firstmonday*, www.firstmonday.dk/issues/issue7_10/press/.

31. Robert Guy Matthews, "Globalization Is Creating U.S. Logistics Jobs," *Wall Street Journal*, March 1, 2004.

32. John McPhee, "Out in the Sort," *New Yorker*, April 18, 2005, 164.

33. Ibid., 167.

34. Ibid., 168.

35. Dani Rodrik, "Sense and Nonsense in the Globalization Debate," *Foreign Policy* (Summer 1997): 31.

Conclusion

1. William T. Stead, *The Americanization of the World* (New York: Garland, 1972), 73, 135.

2. William Dean Howells, *Foregone Conclusions* (1875), 77. Cited in *Oxford English Dictionary*, compact edition, 1:70.

3. Ton von Schaik, "Americanization?" (November 2003), www.uvt.nl/faculteiten/few/economie/schaik/research.html.

4. Theodor Adorno, *Über Jazz. Gesammelte Schriften*, vol. 17 (Frankfurt am Main: Zeitschrift für Sozialforschung, 1935; repr., 1982), 21.

5. Lawrence Grossberg, "Globalization," *Keywords*, www.blackwellpublishing.com/newkeywords/PDFs%20Sample%20Entries%20-%20New%20Keywords/Globalization.pdf.

6. Wolfgang Iser, *The Range of Interpretation* (New York: Columbia University Press, 2000), 5.

7. Ibid., 163.

8. Ryan Bishop and Lillian S. Robinson, *Night Market: Sexual Cultures and the Thai Economic Miracle* (New York: Routledge, 1998), 8.

9. Tomlinson, *Cultural Imperialism*, 90.

10. John Seewer, Associated Press, "Toledo Set to Pioneer Supplier-Run Auto Plant," *Plain Dealer* (Cleveland), August 4, 2004.

Essay on Sources

These abbreviations are used below: *IHT* (*International Herald Tribune*), *NYT* (*New York Times*), *PD* (*Plain Dealer* [Cleveland]), *WSJ* (*Wall Street Journal*).

Chapter 1. "Less Than We Think"

What Are We Talking about When We Talk about Globalization?

The word *habitus* is used by Pierre Bourdieu in *Homo Academicus*, trans. P. Collier (Cambridge: Polity Press, 1988), 194–227, to describe a culturally specific way not only of speaking and doing things, but of seeing, thinking, and categorizing. *Habitus* tends to be "naturalized." It is taken for granted or assimilated into the unconscious, so it becomes the context of action and shared understanding. The passages cited by Fredric Jameson are from his essay in *The Cultures of Globalization* (Durham: Duke University Press, 1998), which he edited with Masao Miyoshi. Thomas Friedman has written often on globalization, most notably *The Lexus and the Olive Tree* (New York: Anchor, 2000) and *The World Is Flat* (New York: Farrar, Straus and Giroux, 2005).

James Brooke wrote about the red bull in "Merrill Lynch's Rush into Japan Ends in Rout," *IHT,* December 15–16, 2001, and Niku's name problems were described in the "Digits" column of *WSJ* on July 22, 2004. Stacy Perman detailed the decline of Levi's in "Levi's Gets the Blues," *Time,* November 17, 1997.

Is English Conquering the World?

Nicholas Wade summarized the paleolinguistic research of Russell D. Gray and Quentin Atkinson, first reported in *Nature,* November 2003, in "A Biological Dig for the Roots of Language," *NYT,* March 15, 2004. The origins of Bahasa Indonesia are detailed in Raymond G. Gordon Jr., ed., *Ethnologue: Languages of the World,* 15th ed. (Dallas: SIL International, 2004), and at http://en.wikipedia

.org/wiki/Indonesian_language. It is based on the Riau Malay spoken in north-east Sumatra, which was the language of only 7 percent of the nation's speakers at independence in 1945, when it officially came into being. It takes vocabulary from Dutch, Portuguese, Chinese, and Arabic, as well as Javanese and other local languages. The articles by Madelaine Drohan and Alan Freeman and Joshua A. Fishman are collected in Patrick O'Meara, Howard D. Mehlinger, and Matthew Krain, eds., *Globalization and the Challenges of a New Century* (Bloomington: Indiana University Press, 2000).

For consistency's sake, the language figures are all from Sidney S. Culbert, University of Washington, *World Almanac*, 1980, 1990, and 1999. The latter was the last year that he supplied these figures. Subsequently most estimates have depended on *Ethnologue,* whose numbers are significantly lower all around but which show the same trends. The discrepancy between these two sources is most obvious in the largest language groups; in Mandarin, for example, the difference between the two sources is 130 million speakers. For Arabic, *World Almanac* stopped listing an aggregate figure after Culbert ceased to report the figures. Newer figures from *Ethnologue* break out Arabic into categories (e.g., Moroccan Arabic, Sudanese Arabic, etc.), which makes it impossible to say in exactly how many overseas nations "Arabic" is spoken or by how many people. My figure is a conservative guess.

For analysis of the U.S. Census figures, see Janny Scott, "Census Shows Big Increase in Foreign-Born U.S. Citizens," *NYT,* February 8, 2002, and Genaro C. Armas, Associated Press, "Non-English Languages Spoken in 1 of 5 Homes," *PD,* October 9, 2003. The use of English in Brazil was reported by the AP, "Brazilian Solon Wants to Give English the Boot," *Daily Yomiuri* (Japan), October 29, 2000. English loanwords appear in the Macmillan English dictionary, available on line at www.macmillandictionary.com/MED-Magazine/april2003/06-language-interference-loan-words.htm. David Puttnam writes about the *benshi* in *Movies and Money* (New York: Knopf, 1998) and Will Ferguson's ESL stories are in *Hokkaido Highway Blues* (Edinburgh: Cannongate,1998).

The Ubiquitous American Film

Key books for understanding the history of U.S. film exports are Kristin Thompson's *Exporting Entertainment: America in the World Film Market, 1907–1934* (London: British Film Institute, 1986), and Thomas Guback, *The International Film Industry* (Bloomington: Indiana University Press, 1969). Much of my understanding of protectionism and runaway production comes from David Puttnam's *Movies and Money.* John Izod offers an excellent overview of how VHS and DVD and overseas box office have changed Hollywood in *Hollywood and the Box Office* (New York: Columbia University Press, 1988).

John Cones, a lawyer, has written the most detailed book on the head-spinning complexity of *The Feature Film Distribution Deal* (Carbondale: Southern Illinois University Press, 1998). Information on most films' box office, rentals, stars, salaries, dubbing and subtitling language, and production locations can be found at www.us.imdb.com. Frederick Wasser was among the first to ask *Is Hollywood America? The Transnationalization of the Film Industry* (Urbana: Institute for Communications Research, 1994), but the *New York Times Magazine*, November 14, 2004 is a special issue on this topic, with valuable articles by Lynn Hirschberg, A. O. Scott, and Manohla Dargis, among others.

For the view that Hollywood film is a "hegemony," see Fredric Jameson in Jameson and Miyoshi, *The Cultures of Globalization*, or Reinhold Wagnleitner, *Coca-Colonization and the Cold War* (Chapel Hill: University of North Carolina Press, 1994). For Mexican film information, see Elisabeth Malkin, "Mexican Film: High Art, Low Budget," *NYT* July 15, 2003. My survey data came from the "Gente" section of Mexico City's *Reforma*, May 20, 1995, as well as "Cartelera," *La Jornada*, May 15, 1995, and "Gente," *Reforma*, May 20, 1995. In Japan I used *Kansai Time-Out*, *Yomiuri Daily*, and *Japan Times*. Other data are from Chang-Ran Kim, "Hopeful Multiplexes Spin Shaky Web, Bank on Hits," in *Japan Times*, May 24, 2002. Figures on the earnings of Japanese films are from *Japanese Film: 2002* (Tokyo: UniJapan Film, 2002) and available at www.asianfilms.org/japan/yearreview.html. For French film statistics, see "French Film Industry Fears Its Worst-Case Scenario," *IHT*, December 27, 2001. For Russian and Nigerian figures, see "Russian Film Industry Battles Hollywood Goliath," *China Daily*, June 18, 2002; "Red to Noir: Russian Film's Gritty Comeback," *WSJ*, October 14, 2003; and L. Riding, "Film-Makers," *NYT*, February 5, 2003. Sean Park gives the South Korean figures in "Coming to a Theater near You?" *WSJ*, October 31, 2003. Film in Cleveland described by the *PD*, July 4, 2003, as well as by Thomas Ott, "Reel Success Is Hard Work," *PD*, December 31, 2002. For the overseas star power of Leonardo DiCaprio, see John Lippman, "Bombs—Away!" *WSJ*, November 19, 2004. On Italian dubbing, see Alessandra Stanley, "Italians Can't Believe Ears: Movies Lose Their Voices," *IHT*, September 17, 1998. I want to thank Rob Spadoni for reading this section and suggesting changes.

American Television and the Rise of Local Programming

A good place to begin reading is *Television: An International History*, edited by Anthony Smith (Oxford: Oxford University Press, 1998); it contains William Boddy's essay on early TV in the United States and Richard Paterson's on foreign viewing habits in drama (including Africa's). Peter L. Berger's essay "Four Faces of Global Culture," in O'Meara et al., *Globalization*, and Frederick Wasser's *Is Hollywood America?* contain insights on international television markets. For drama content analysis, see Divina Frau-Meigs, the leading European scholar, whose *Médiamorphoses américaines dans un espace privé unique au monde* (Paris: Economica, 2001) is available in the United States. Her early work is summarized in "The Cultural Impact of American Television Fiction in Europe: Transfer of

Imaginary Worlds or Cultural Compatibility?" a paper presented at the European American Studies Association, Warsaw, Poland, March 21–24, 1997.

The major profile of Televisa is "Televisa tend ses chaines sur le Mexique," *Libération* (Paris), June 4–5, 1994, but the Mexican television industry has been detailed by Elizabeth Malkin, "Mexico Media Mogul Follows the Money," *NYT*, February 27, 2004, and Cecilia Bouleau, "Madison Ave. South," *News* (Mexico City), May 22, 1995. The *Los Angeles Times* follows Mexican television with interest: see viewership statistics published June 22, 1998. My own observation of Televisa studios and Mexico City television used information from *News* (Mexico City), May 22, 1995, and "Cultura," *Reforma,* May 20, 1994.

The *Wall Street Journal* follows television as an industry; see Bill Spindle's "On TV in Syria: Satire, Corruption, Religious Tensions," May 5, 2005, and Brookes Barnes and Miriam Jordan, "Big Four TV Networks Get a Wake-Up Call—in Spanish," May 2, 2005. The *New York Times* also keeps tabs on the industry: "Slower Growth in Cable," January 26, 2004, detailed the Chinese market, and Frank Bruni chronicled "A Search for Girls, Girls, Girls around Italy's Dial," September 9, 2002. Marc Lacey wrote "Reality TV Rivets Africa, to the Churches' Dismay," reprinted in *IHT,* September 3, 2003.

Costs of launching a television station were explained to me by Pape Gaye, Foreign Ministry, Dakar, Senegal, May 14, 2000, and Saad Asswailim, Embassy of Saudi Aradia, Washington, D.C., August 12, 2004.

There are many good books on Disney. Richard Schickel's *The Disney Version: The Life, Times, Art & Commerce of Walt Disney* (Chicago: Dee, 1997) is perhaps the standard, but see also Karal Ann Marling (no relation), ed., *Designing Disney's Theme Parks: The Architecture of Reassurance* (Paris: Flammarion, 1998). Ariel Dorfman and Armand Mattelart's *How to Read Donald Duck* (New York: International General, 1971) received a new introduction by translator David Kunzle in 1995 that demonstrates how widely outsourced comics are. For more information on the world of cartooning, see Michael Switow, "Philippines Is Cartoon Capital," *PD,* October 10, 1996.

The McDonald's Brouhaha

The best recent treatment of McDonald's is Stephen Drucker, "Who Is the Best Restaurateur in America?" *New York Times Magazine,* March 10, 1996. Much information is at McDonald's websites, www.mcdonalds.com/corporate/franchise/faq/ and www.licenseenews.com/news/news182.htm. See also James L. Watson's *Golden Arches East: McDonald's in East Asia.* (Stanford: Stanford University Press, 1997), which contains a good overview by Emiko Ohnuki-Tierney. George Ritzer and Elizabeth L. Malone's claim that "In Korea (and Japan) the individualism of a meal at McDonald's threatens the commensality of eating rice that is cooked in a common pot and of sharing side dishes" is in *American Studies* 41, nos. 2–3 (2000): 107. For McDonald's in Russia, see Erin E. Arvedlund, "McDonald's Commands a Real Estate Empire in Russia," *NYT,* March 17, 2005. On New York City's Chinese restaurants, see Michael Spector, "Fashion Cafeteria," *New Yorker,* September 27, 2004.

What about the Internet?

Most information on computer use is available, not surprisingly, on line. The Nielsen/NetRatings are at www.nielsen-netratings.com, while the NUA surveys are at www.nua.ie/surveys/how_many_online/index.html. VeriSign is at www .verisign.com/nds/naming/idn/value/market_research.

Matthew Zook is a one-man army when it concerns Internet use. His most recent book is *The Geography of the Internet Industry* (London: Blackwell, 2005), but he has numerous papers that spell out his findings, some available at his website, www.zooknic.com. For his domain name research, go to www.zooknic.com/ Domains/international.html.

The *New York Times* has followed the story of computers and "dying languages" closely. See Nicholas Wood, "In the Old Dialect, a Balkan Region Regains Its Identity," February 24, 2005, and Marc Lacey, "Using a New Language in Africa to Save Dying Ones," November 12, 2004. Wayne Arnold reported on "Malaysia's Internet Road Show," August 23, 2001. Gregory Beals highlighted broadband gaming in "All the World's a Game," *Newsweek,* International edition, July 9, 2001, and *Time* detailed the South Korean government applications in its August 20, 2001, issue.

Do American Companies Dominate the World Economy?

Fortune's list of the world's largest companies for 2003, ranked by revenue, appeared at www.fortune.com/fortune500/company/top500. The UN's Commission on Trade and Development reports are found at www.unctad.org/ Templates/Page.asp?intItemID=1485&lang=1. The *Wall Street Journal* reports of concern were Stephen Power, "GM Plans Major Overhaul of Its Business in Europe," June 17, 2004; Ann Zimmerman and Martin Fackler, "Wal-Mart's Foray into Japan Spurs a Retail Upheaval," September 19, 2003; Dan Bilefsky, "SAB-Miller Still Is Flat in U.S., Foamy in Europe," September 30, 2003; and "Cingular Deal Will Reshape Industry," February 19, 2004. On the geographic distribution of TNCs, see George Melloan, "Feeling the Muscles of the Multinationals," *WSJ,* January 6, 2004, and *Global Inc.* by Henry Bruner and Gabel Medard (New York: Free Press, 2003). On Citibank in Japan, see Andrew Morse and Mitchell Pacelle, "Japan Orders Citibank to Halt Private Banking," *WSJ,* September 20, 2004, while the *Economist,* "Flying on Empty," May 21, 2005, reported on the woes of the U.S. airlines.

The *New York Times* business pages gave figures on share of the auto industry in its September 4, 2003, issue: GM's 2002 market share was 25.5 percent, Toyota's was 14.3 percent, Ford's (including Volvo and Land Rover) was 13.5 percent,

Honda's was 13 percent, and DaimlerChrysler's was 6.3 percent. The best-selling vehicles were (1) Ford F-series light trucks, (2) Chevy Silverado light trucks, (3) Honda Accord, (4) Toyota Camry, (5) Dodge Ram light trucks, (6) Ford Explorer, and (7) Toyota Corolla. All Japanese models increased sales over the previous year, but only the Silverado increased sales among American vehicles. On Citgo and Venezuela, see Simon Romero and Alexei Rarrionuevo, "The Troubled Oil Company," *NYT,* April 20, 2005. Sherri Day and Tony Smith wrote "Interbrew Said to Be Near Deal for Brazil Brewer," *NYT,* March 2, 2004.

For GM's role in Mexico, see Charles W. McMillion, *Assessing NAFTA* (Washington, D.C.: MBG Information Services, 1999). GM and Ford are among the largest companies in Mexico: the value of auto parts exported from Mexico to the United States rose from $7.4 billion in 1993 to $14.5 billion in 1998. See also James Brooke, "Japan-Mexico Free Trade Talks Falter," *NYT,* October 17, 2003.

Bloomberg News reported "New Releases of Music Lift PolyGram's Profit 95%," *IHT,* October 22, 1998; and Thomas W. Gerdel reported "Reinventing the Tire," *PD,* March 20, 2005. Chinese shoe production in "Province Makes 30% of the World's Shoes," *Guangzhou Morning Post,* June 14, 2002. On Carrefour's Chinese employees, see Mark O'Neill, "Carrefour Reveals Official Ties a Bind," *South China Morning Post,* November 30, 2004. On foreign employees in Prague, see Keither Damsell, "More Foreign Firms Seeking Czech Employees," *Prague Post,* April 6–12, 1994. P&G inventory information is from interview with Naoko Yamaguchi, Procter & Gamble programmer, Yobe, Japan, May 5, 2000.

Chapter 2. The Resistance of the Local

Language

Dusan Kecmanovic has given us one of the fullest and most convincing accounts of why ethnic and nationalistic thinking will not go away, in *The Mass Psychology of Ethnonationalism* (New York: Plenum Press, 1996). Two studies used extensively in this section of the book, which incidentally highlight the importance of local culture, are Geert Hofstede, *Cultures and Organizations: Software of the Mind* (New York: McGraw-Hill, 1997), and Ronald Inglehart, *Modernization and Postmodernization: Cultural, Economic, and Political Change in 43 Societies* (Princeton: Princeton University Press, 1997). For reportage on Japan, no one exceeds Norimitsu Onishi, "Japanese Workers Told from on High: Drop the Formality," *NYT,* October 30, 2003. Onishi recounts the famous 1975 case of a junior employee in Tokyo who was beaten to death when he refused to use the proper *keigo* with a senior colleague in a bar. For the changeover in Chinese, see David W. Chen "egap a nrut .S.U. ni srepapswen esenihC," *IHT,* March 28, 2002. The Chuvashi information is in Steven Lee Myers, "Russian Republics Assail Putin's Plan to Curb Autonomy," *NYT,* October 4, 2004.

Communicative Distance

On the *hikikomori* phenomenon, see Kathryn Tolbert, "They Won't Leave Their Rooms," *IHT,* May 31, 2002. Simon Rowe wrote an amusing review of capsule hotels, "Sure It's Tiny, It Has No View, but Hey, It's Home," *IHT,* March 16–17, 2002. The information on which nationalities travel came from "Euro: Single Currency," *IHT,* December 11, 2001, and Crispian Balmer wrote about Italians living at home: "30 and Still Living at Home," *IHT,* April 6, 2002.

Anthropologist Edward Hall's system of proxemics, detailed in *Beyond Culture* (Garden City, N.Y.: Anchor Press, 1976), set out the terms *low context* and *high context*. Kurt Lewin, the "father" of group dynamics at MIT, had in the 1950s dubbed these *U-type* and *G-type*, and later developed "field theory." As the terms suggest, both are spatially oriented systems.

Food

Pierre Bourdieu's *Distinction: A Social Critique of the Judgement of Taste* (Cambridge, Mass.: Harvard University Press, 1984) influenced this chapter, obviously, along with Sidney Mintz, *Tasting Food, Tasting Freedom: Excursions into Eating, Culture, and the Past* (Boston: Beacon, 1996). John Dower's *Embracing Defeat* (New York: W. W. Norton, 1999) gives much useful context to modern Japanese eating habits. The information on breast-feeding comes from Lizette Alvarez, "Norway Leads Industrial Nations Back to Breast-Feeding," *NYT,* October 21, 2003, and Miriam Jordan, "Nestlé Markets Baby Formula to Hispanic Mothers in U.S.," *WSJ,* March 4, 2004, which quotes from an Abbott Labs study showing that in the United States 46 percent of Asian mothers, 36 percent of Caucasian mothers, 32 percent of Hispanic mothers, and 19 percent of black mothers were still breast-feeding their babies at six months. I also interviewed many mothers: my thanks to Akiko Hasegawa, December 2, 2000, and Naoko Kumise, June 24, 2004, in Japan, and Anne Luyot, Jacqueline Miquet, and Beatrice Laurent in France, November and December, 2001. On U.S. infant diets, see Mintz and also T. A. Badget, "Bad Eating Habits Start before 2, Study Finds," *PD,* October 26, 2003.

The failure of low-fat products is described by Bruce Horowitz, "Low-Fat Industry Loses Out as Consumers Favor Flavor," *USA Today,* October 15, 2001, which says that sales of low-fat products fell from 10 to 50 percent in every category between 1995 and 2001. Warren Hoge wrote about marmite: "It's Revolting—and Sublime," *IHT,* January 25–26, 2002. For details on Starbucks's overseas adaptations, read Susan Leung, "Overseas, Starbucks Learns as It Expands," *WSJ,* reprinted in the *Jerusalem Post,* December 16, 2003. For Chinese food shopping habits and cooling and warming flavors, see Geoffrey A. Fowler and Ramin Setoodeh, "Outsiders Get Smarter about China's Tastes," *WSJ,* August 5, 2004.

Gender

Geert Hofstede's study, *Cultures and Organizations,* based on interviews of 116,000 IBM employees in seventy-two countries, is particularly suited to sort out gender roles in postmodernity. His basic gender tool is a masculinity-femininity index. At the masculine end we find Japan and Austria. The United States is in the middle between Ecuador and Australia, while the Scandinavian countries are the most feminine. "While all industrial societies over the past decades have shown a gradual increase in female participation in the work force," writes Hofstede, "ambitious women are more frequently found in masculine rather than feminine societies" (95–96). "Gender relations have similarly become much more complicated as resort to a female labor force has become much more widespread," according to the geographer David Harvey in *The Condition of Postmodernity* (London: Blackwell, 1989), 192. This complexity, with its varying regional roots, has led to greater heterogeneity in gender roles.

Thai prostitution is the subject of much hyperbolic academic concern: Ryan Bishop and Lillian S. Robinson have weighed in with *Night Market: Sexual Cultures and the Thai Economic Miracle* (New York: Routledge, 1998). Joanne Nagel, in "States of Arousal/Fantasy Islands: Race, Sex, and Romance in the Global Economy of Desire," *American Studies* 41, nos. 2–3 (2000), writes that "These patterns [of sex tourism] have been recently reinforced by World Bank and International Monetary Fund economic restructuring programs under which both migrating and local 'women are increasingly more active in informal economies, which includes sex industries'" (169). While she seems to indicate that the World Bank or IMF supports the Thai sex trade, she is in fact quoting from a highly politicized tract on Caribbean migration by Kamala Kempardoo. For more on Asian women and body styles, see Cris Prystay and Geoffrey A. Fowler, "For Asian Women, Weight Loss Rule 1 Is Skip the Gym," *WSJ*, October 10, 2003. The Iraq situation is detailed in Robert F. Smith, "In Jeans or Veils, Iraqi Women Are Split on New Political Power," *NYT*, April 13, 2005. Karen Kelsky's research is in "Flirting with the Foreign: Interracial Sex in Japan's 'International Age,'" in Rob Wilson and Wimal Dissanayake, eds., *Global/Local: Cultural Production and the Transnational Imaginary* (Durham: Duke University Press, 1996), 173–12.

Education

Roberta Wollons introduced me to the inner workings of the educational system and much of the rest of Japan, for which I owe her great thanks. Her book *Kindergartens and Cultures: The Global Diffusion of an Idea* (New Haven: Yale University Press, 2000) is an excellent starting place for reading. My thanks also to Phillip McClellen and Dawn Grimes-McClellen, teachers and neighbors in Japan, for their insights. Joseph Tobin's account of Hiroki appears in Merry L. White and Sylvan Barnett, eds., *Comparing Cultures: Readings on Contemporary Japan for American Writers* (Boston: St. Martin's Press, 1995), 89–94, as does Susan Chira's essay on Koreans in the Japanese education system (428–30). See also White's "Children and Families: Reflections on the 'Crisis' in Japanese Childrearing Today," in Hidetada Shimizu and Robert A. LeVine, eds., *Japanese Frames of Mind: Cultural Perspectives on Human Development* (Cambridge: Cambridge University Press, 2001). Shintaro Ishihara, governor of Tokyo prefecture and prominent opponent of the six-day school week, called it a "classic ploy of central government to keep everything under rigid control" in an interview with the *Asahi Shimbun,* May 5, 2002. Thomas P. Rohlen is an authority on Japanese education: see Rohlen and Gerald K. LeTendre, eds., *Teaching and Learning in Japan* (Cambridge: Cambridge University Press, 1996), and Rohlen's *Japan's High Schools* (Berkeley: University of California Press, 1983). Also good is William K. Cummings and Philip G. Altbach, eds., *The Challenge of Eastern Asian Education: Implications for America* (Albany: State University of New York Press, 1997).

The French attitude toward education makes more sense seen through the lens of Ronald Inglehart's emplotment of France and Japan on the axes of the secular versus the traditional in *Modernization and Postmodernization.* The Japanese are more oriented to secular rationalism.

Work

The statistics on productivity are from the International Labor Organization and reported by Steven Greenhouse, "U.S. Growth Industry: Workdays," *IHT,* September 3, 2001. Because vacation periods and holidays vary, it's difficult to put the statistics into a firm comparative framework. The comparison of AOL programmers around the world is originally from the *NYT,* reprinted in Chris Oakes, "Programmers Share a Code the World Over," *IHT,* March 15, 2002.

Land Use

There are many on-line databases covering land use in specific nations; I used demographia.com for my figures on Japan and other nations. A good start for reading on U.S. land use is Marion Clawson, *The Land System of the United States: An Introduction to the History and Practice of Land Use and Land Tenure* (Lincoln: University of Nebraska Press, 1968). Two other classic studies are Jane Jacobs's *The Death and Life of Great American Cities* (New York: Vintage, 1992) and Lewis Mumford, *The City in History: Its Origins, Its Transformations, and Its Prospects* (Boston: Harvest Books, 1968). More recent are Alex Marshall's *How Cities Work: Suburbs, Sprawl, and the Roads Not Taken* (Austin: University of Texas Press, 2001) and Peter Katz Alex, *The New Urbanism: Toward an Architecture of Community* (New York: McGraw-Hill, 1993). On recent U.S. land use, see the "Class Matters" series in *NYT,* May 25–June 3, 2005, particularly Peter T. Kilborn's "The Five Bedroom, Six-Figure Rootless Life" on June 1, 2005. The figures on Americans' auto spending are found in Jeffrey Ball, "For Many Low-Income Workers, High Gasoline Prices Take a Toll," *WSJ*, July 12, 2003. Alex Kerr's books on Japan are *Lost Japan* (Oakland: Lonely Planet, 1996) and *Dogs and Demons: Tales from the Dark Side of Japan* (New York: Hill and Wang, 2001). See also James Fallows's *Looking at the Sun: The Rise of the New East Asian Economic and Political System* (New York: Pantheon Books, 1995) and John Dower's *Embracing Defeat: Japan in the Wake of World War II* (New York: Norton, 2000). David Yetman, "*Ejidos*, Land Sales, and Free Trade in Northwest Mexico," *American Studies* 41, nos. 2–3 (2000): 211–34, is the source of material on Mexico, but see also Wayne Cornelius and David Myhre's contributions to *The Transformation of Rural Mexico: Transforming the Ejido Secto*r (San Diego: Center for U.S.-Mexican Studies, University of California at San Diego, 1998), as well as Alberto Burquez's "Twenty-Seven: A Case Study in Ejido Privatization in Mexico," *Journal of Anthropological Research* 54, no. 1 (1998): 73–95, and John Womack's *Zapata and the Mexican Revolution* (New York: Vintage, 1970). On French urban planning, I found Gilles Blieck's *Les Enceintes Urbaines, XIII–XVI siècle* (Nice: Editions du CTHS, 1999) very helpful.

Tribalism

Racism shows up in other Japanese sports besides baseball. Until the ascent in the 1990s of Akebono, a Hawaiian, no foreigner had ever reached the top group of *sumo*. By 2001 there were several Mongolians in the second *juryo* group, as well as—gasp!—a black American named Henry Miller. Miller's heavily muscled body (he trained with weights) led newspaper commentators to allege that *gaijin*— who are "genetically different"—would ruin *sumo* by introducing steroids. They urged a ban.

Kofi Annan's critique of African racism is in "The Politics of Globalization," an address presented at Harvard University, Cambridge, Mass., September 17, 1998. Reprinted in O'Meara et al., *Globalization*, 125–30. Senegal's president Abdoulaye Wade quoted by Joseph R. Gregory, "In Mali, a Fragile Democracy," *IHT*, April 26, 2004. The Hem and Lendus conflict in Congo detailed by Somini Sengupta, "French Soldiers Arrive in Congo with a Tough Mission," *NYT*, June 7, 2003. For more on Mozambique and Zimbabwe, see Michael Wines, "We Welcome You to Lush Zimbabwe! Your Wallet, Please!" *NYT*, November 19, 2003. On Rwanda, read Gil Courtemanche's *A Sunday at the Pool in Kigali* (New York: Knopf, 2003). The "logo did it" view of African racism is George Packer's "Letter from the Ivory Coast: Gangsta War," *New Yorker*, November 3, 2003. Twenty years earlier V. S. Naipaul had depicted a decline into tribalism in "The Crocodiles of Yamoussoukro," *New Yorker*, May 23, 1983.

Seymour Hersh revealed Israeli aid to the Kurds in "Plan B," *New Yorker*, June 28, 2004. On Spanish anti-Moroccan violence, see Marlise Simons, "Resenting African Workers, Spaniards Attack," *NYT*, February 12, 2000. On Austria, see Steven Erlanger, "How the Viennese Rushed to 'Aryanize' Property," *IHT*, March 8, 2002. On the problems of German companies in foreign acquisitions, see

Matthew Karnitschnig, "A Newspaper War in Poland Tests Europe's Barriers," *WSJ*, October 20, 2004. Violence in India detailed in Rajiv Chandrasekaran's "'Both Sides Were at Fault,' in India," *IHT*, March 7, 2003, and on caste, see Joseph Berger, "Family Ties and Entanglements of Caste," *NYT*, October 25, 2004, and Katharine Boo, "The Best Job in Town," *New Yorker*, July 5, 2004.

Tonga's problems came to light in "Royal Jester's Date in Court No Joke," *South China Morning Post*, June 6, 2002. On Mexican intrafamilial racism, see Jacqueline Fortes de Left, "Racism in Mexico: Cultural Roots and Clinical Interventions," in *Family Process* 41 (Winter 2002): 619. On Brazilian slavery, see Larry Rohter, "Trapped Like Slaves on Brazilian Ranches," *IHT*, March 26, 2002, while Anthony Faiola reported on *The Simpsons* in "The Cartoon That Riled Rio," *IHT*, April 18, 2002.

Corruption

The Corruption Perceptions Index is available at www.transparency.org. The "Data and Research" section, for 2002, pp. 224ff., identified fourteen sources for its data: World Bank Group, World Economic Forum, World Business Environment Survey, Institute of Management Development, PricewaterhouseCoopers, Political and Economic Risk Consultancy, The Economist Intelligence Unit, and Freedom House. Other studies mentioned were authored by the Control Risks Group; Dow Jones Sustainability Group Index; The Conference Board; Transparency International; World Bank PREM Network; Latinobarometer; Afrobarometer; International Crime Victims Survey; New Europe Barometer; and several other organizations.

The Bribe Payers Index is conducted by Gallup International for Transparency International; it is contained in the larger Corruption Perceptions Index, pp. 237–39. The study on the "Amount Corruption Adds to Bureaucrats' Salaries" is contained in the Global Corruption Report, pp. 296–98, www.globalcorruptionreport .org.

Stephan Faris reported on oil companies in Nigeria in "Oil Giant Could Do Better in Nigeria," *Fortune*, October 1, 2001. For more on Argentina's fleecing of bondholders, see Matt Moffet, "After Huge Default, Argentina Squeezes Small Bondholders," *WSJ*, January 14, 2004. Thomas Friedman in *The Lexus and the Olive Tree* has many revealing anecdotes on corruption. For more on the embeddedness of corruption in Southeast Asia, see Cas W. Vroom, *Indonesia and the West: As Essay on Cultural Differences in Organization and Management* (Jakarta: Catholic University Press, 1981), and Emmanuel Todd, *The Explanation of Ideology: Family Structures and Social System* (Oxford: Blackwell, 1985).

Smuggling and Counterfeiting

Timothy W. Ryback's *Rock around the Bloc* (New York: Oxford University Press, 1990) is an enjoyable way to start reading. Peter Jaszi and Martha Woodmansee have collected a variety of academic approaches to the problem in *The Construction of Authorship: Textual Appropriation in Law and Literature* (Durham: Duke University Press, 1994).

Tim Phillips, in "A Web of Corporate Cops and Robbers," *IHT* Biztech, March 28, 2003, lists the nations using pirated business software, a role call of the underdeveloped world, led by Indonesia (89%), Ukraine (89%), and Russia (88%), but it is worth noting that 25 percent of the business software used in North America and 35 percent of that used in Europe is also pirated

Taxes

The tax compliance rates are available from Transparency International and are cited in "Global Survey Cites Creeping Corruption," *NYT,* October 8, 2003. The Lebanese tax mess is described by Salim Yassine, "In Beirut, a Taxing Situation," *IHT,* February 7, 2002.

The Resistance of the Local

On Japanese and American savings rates, see Justin Doebele, "The El Dorado of Japan," *Forbes Global,* October 15, 2001. Carol Hymowitz explains European deodorant use in "European Executives Give Some Advice on Crossing Borders," *WSJ,* December 2, 2003. On Whirlpool's different washing machines, see Miriam Jordan and Jonathan Karp, "Machines for the Masses," *WSJ,* December 9, 2003.

Chapter 3. *"More Than We Know"*

ATMs

For information on Don Wetzel and his colleagues' invention of the ATM, see inventors.about.com/library/inventors/blatm.htm. Other claimants are Barclay's Bank of England, which claims to have had the first "cash dispenser" in 1967, but its machine simply exchanged bank script for ten-pound notes, and the OMRON-Taiyo Corp. of Beppu, Japan, which claims to have "developed and delivered" the "first in the world" cash dispenser in 1971. I have found no photos or other substantiation for this claim either. Foreign use of ATMs will probably continue to rise even as U.S. use appears to have crested. According to *ATM & Debit News,* 65 percent of American households used the machines in 2000, but only 57 percent in 2003, as they were replaced by debit card purchases. See Sasha Talcott, "U.S. Moves Closer to Cashless Economy as ATM Use Falls," *PD,* July 22, 2004. Information on number of ATMs over the years comes from "world situation today" at www.rbrldn.demon.co.uk/history.htm. The information is taken from *The Global ATM Market to 2004,* published by Retail Banking Research Ltd. Also see http://military.bankrate.com/mtry/green/atm/atm4a.asp?prodtype=bank.

Sales figures are closely guarded, but some information is available at company websites, such as www.tritonatm.com. Craig Karmin is the source of the information on South Korean and Chinese ATM and credit card use: "China Prods Its Consumers to Use Plastic," *WSJ*, December 4, 2003. Eric Bellman wrote the wonderful story on ATMs in India: "In India, Small Savers Mean Big Profit," *WSJ*, July 7, 2004. For more on the Hong Kong and Singapore stored value cards, see Evan Ramstad, "Hong Kong's Electronic-Money Card is Hit," *WSJ*, February 10, 2004.

The Money Market

Despite its metaphors, Thomas Friedman's chapter on money markets in *The Lexus and the Olive Tree* is still the best place to begin reading on this subject. My account relies on many of his details. For more complex analysis, see Saskia Sassen, *Globalization and Its Discontents* (New York: New Press, 1999). On the failure of "development economics," see William W. Lewis, *The Power of Productivity* (Chicago: University of Chicago Press, 2005), especially 135–228. My thanks to my colleague Asim Erdilik at the Weatherhead School, CWRU, for reading this section and suggesting improvements. For more on the way foreigners' investments in the United States have kept interest rates low, see Alan Murray's column, "Rates Lay Bare Greenspan Challenge," *WSJ*, June 1, 2005. The *Economist* has written extensively on foreign direct investment: see December 25, 1997 (cited by Friedman, 121) and the argument of April 16, 2005, for flatter tax rates, as well as "The New Kings of Capitalism," *Economist*, November 27, 2004.

Flexible Manufacturing

Fifteen years ago, this type of foreign investment accounted for only 0.4 percent of the gross domestic product of underdeveloped nations. By 2000 it had quintupled to 2 percent, and it appears set to rise to 5 percent by 2010 according to the UN Commission on Foreign Trade and Development. For more on Toyota, read Norihiko Shirouzu and Sebastian Moffett, "Toyota Closes in on GM," *WSJ*, August 4, 2004. Information on Flextronics can be found at www.tdctrade.com/imn/04062404/consumerele032.htm and in Jeffrey M. O'Brien's "The Making of the Xbox," *Wired*, issue 9.11, November, 2001.

Franchising

The word *franchise* dates to the twelfth century and derives from the French *franc* or *franche*, to be free, especially in judgments moral and metaphysical (*Petit Robert: Dictionnaire de la langue française* [Paris: Le Robert, 1986], 822–23). Historic examples in this section are from Robert Rosenberg and Madelon Bedell, *Profits from Franchising* (New York: McGraw-Hill, 1969); Robert M. Dias and Stanley I. Gernick, *Franchising: The Investors' Complete Handbook* (New York: Hastings House, 1969); and Michael R. Czinkota, Ilkka A. Ronkainen, and John J. Tarrant, *The Global Marketing Imperative* (Lincolnwood, Ill.: NTC Business Books, 1996).

On McDonald's adoption of credit card sales, see Jathon Sapsford, "As Cash Fades, America Becomes a Plastic Nation," *WSJ*, July 23, 2004. Joel Millman and Ann Zimmerman tell Roxana Orellana's story in "'Repats' Help Payless Shoes Branch Out in Latin America," *WSJ*, December 24, 2003; Australian franchising by Julie Bennet, "Foreign Franchise Concepts Find Growth Opportunities in U.S. Market," *WSJ*, March 8, 2004; and Kumon Learning Centers by Suein Hwang, "Pre-K Prep: How Young Is Too Young for Tutoring?" *WSJ*, October 13, 2004.

Airfreight

The effects of the 9/11 air traffic delays were reported by Mark Landler, "Air Cargo Delays Have a Ripple Effect," *IHT*, September 15, 2001, and James Brooke, "Attacks Cause a Ripple through Japan's Economy," *IHT*, September 21, 2001.

On the history of airfreight in the United States, see Richard Malkin, "An Air Cargo Century," *Air Cargo World On-Line* (2000), www.aircargoworld.com/archives/feat1jan00.htm. The FedEx website is at www.fedex.com/us/about/express/history.html and Magic Millions-IRT at www.magicmillions.com.au/air_freight.html. On Apopka greenhouses, see Florida Agricultural Statistics Service reports, on line at www.nass.usda.gov/fl/hort/f&fcg98.htm.

On Chinese air cargo services, see Amy Schatz and Rick Brooks, "Polar Air Cargo Wins Rights to U.S.-China Service," *WSJ*, September 7, 2004, www.aircargoworld.com/archives/feat2dec99.htm, and Cathay Pacific, *Freighter Flight Schedules* (Hong Kong, 2004). For per mile cost figures and much other interesting data, read Hendrik Tennekes, *The Simple Science of Flight: From Insects to Jumbo Jets* (Boston: MIT Press, 1996).

Containerized Freight

Statistics on New York–New Jersey ports and the *Hyundai Glory* come from Eric Lipton's extensive series in *NYT*, "New York Port Hums Again, with Asian Trade," November 22, 2004, and "Beneath the Harbor, It's Dig or Else," November 23, 2004. Information on the SS *Elizabethport* and Oakland container port is at www.portofoakland.com/about/history.html. On Hong Kong, see http://geography.miningco.com/library/faq/blqzenclosedbuilding.htm. The *South China Morning Post* follows this industry closely; see "Overstretched Kmart Files for Bankruptcy," January 23, 2002. To learn about the Maersk line, go to www.maerskline.com/vessels. Tokyo's Shinogawa Terminal details at www.kouwan.metro.tokyo.jp/kowane/rekisi/rekisie.html. Los Angeles and Long Beach port capacities from the World Bank Study by Paul O. Roberts, "Trends in Inter-Regional Goods Movement," at www.worldbank.org/html/fpd/transport/ports/trf_docs/trends.pdf. On the ILU, start with www.metrans.org/Research/draft_reports/AR_04-02_final_draft.htm. For the Los Angeles and Long Beach port battles, see Daniel Machalaba, "U.S. Ports Hit a Storm," *WSJ*, March 10, 2004. On Detroit's problems, see Christopher M. Singer, "Neighborhood vs. Business,"

Detroit News, January 9, 2001 (on line at http://detnews.com/2001/detroit/0109/19/s06-296861.htm). Vietnam's new ports described at http://vietnamnews.vnagency.com.vn/2001-02/19/Stories/04.htm. The World Bank's report on roads and walking is at www.worldbank.org/transport/roads/pov&sa.htm.

Bar Codes

Two key resources are Tony Seideman's "Bar Codes Sweep the World," *American Heritage of Invention and Technology* 8, no. 4 (Spring 1993), which is cited at: www.inventionandtechnology.com/xml/1993/4/it_1993_4_toc.xml, and Roger C. Palmer, *The Bar Code Book: Reading, Printing, and Specification of Bar Code Symbols* (Peterborough, N.H.: Helmers, 1995). For more on warehouse space optimization, www.syware.com/reflib/hanimp.htm, and on RFIDs, see Chris Seper, "Radio Shipping Tags, Ready or Not!" *PD,* July 28, 2004.

Computing

For more on Google, see Steven Levy, "All Eyes on Google," *Newsweek,* March 29, 2004, and Kevin J. Delaney and Andres Cala, "France Mobilizes, Seeks European Allies to Fend Off Google," *WSJ,* May 12, 2005, and John Markoff and Edward Wyatt, "Google Is Adding Major Libraries to Its Database," *NYT,* December 14, 2004.

Research on negative opinions at eBay reported on "Nobel Prize Winners in Economics Announced," *Morning Edition,* National Public Radio, October 13, 2004. See also Nick Wingfield and Jeffrey A. Trachtenberg, "Amazon Prods Reviewers to Stop Hiding behind Fake Names," *WSJ,* July 28, 2004. On Yahoo's auction business in Japan, see www.businessweek.com/magazine/content/01_23/b3735139.htm.

For research on the Internet in China, see Emily Parker, "China vs. the Internet," *WSJ,* May 3, 2004, and Julianna Liu, "China's Web Search Engines Set to Take on Google," http://Forbes.com/technology,newswire,2004/03/10/rtr129281. On Sina.com, see Daniel Sneider, "Internet China Is a 'Nation' Resisting Government Control," KansasCity.com, March 25, 2004. The three sites where I found product reviews in Chinese, with the aid of my colleague Tsiu Liang, were http://newegg.people.com/cn/products/ProductReview.aspx?SysNo=4122; www.yongle.com.cn/service/forum/viewforum.asp?root=4; and my.cnd.org/modules/newbb/viewtopic.php?mode=viewtopic&topic_id=.

For more on Chinese Internet censorship, see Jonathan Zittrain and Benjamin Edelman, "Empirical Analysis of Internet Filtering in China," www.cyber.law.harvard.edu/filtering/china/. The spam situation was described by Mei Fong, "The Spam-China Link," *WSJ,* July 21, 2004.

Logistics

For more on Procter & Gamble, see Davis Dyer, Frederick Dalzell, and Rowena Olegario, *Rising Tide* (Boston: Harvard Business School Press, 2004). Gabriel Kahn detailed the disappearing letter of credit in "Financing Goes Just-in-Time," *WSJ*, June 4, 2004. John McPhee's article on UPS—"Out in the Sort," *New Yorker*, April 18, 2005—is by far the most interesting and humanizing profile of logistics and of this company.

Conclusion

William T. Stead's book, *The Americanization of the World* (New York: Garland, 1972), is available in reprint. For historic definitions of *Americanization*, see the *Oxford English Dictionary*, compact edition (1971), 1:70. Gordon Wood's book is *The Americanization of Benjamin Franklin* (New York: Penguin, 2004). Luigi Barzini wrote insightfully (and humorously) about differences between the United States and Europe in *Oh America, When You and I Were Young* (New York: Viking, 1985) and *The Europeans* (New York: Penguin, 1984). Paul Fussell's observations are in *The Great War and Modern Memory* (New York: Oxford University Press, 2000) and *Wartime: Understanding and Behavior in the Second World War* (New York: Oxford University Press, 1990).

Information on foreign companies in Ohio is from Stephen Koff, "Honda Workers Are Example of Flip Side of Job Outsourcing," *PD*, May 3, 2004; WVIZ-WCPN television and radio, 8:30 a.m., April 28, 2004; and John Seewer, Associated Press, "Toledo Set to Pioneer Supplier-Run Auto Plant," *PD*, August 4, 2004.

I want to thank my colleague Gary Stonum for reading and critiquing this book. He's been generous to me over and over again.

Index

search engines, 183–88
Seideman, Tony, 179–80
Sesia, Davide, 85
7-Eleven, 2, 151, 201
sex tourism, 98–99
Schindler's List, 36–37
Silver, Bernard, 179–81
Slavenburg Bank, 24
Smith, Frederick W., 169–70
Sodexho, 59, 76, 201
Sohio, 3
Solectron, 161
Sonnenborn, Harry, 163
Sony, 25, 69, 78
Soros, George, 155
South Korea: container ships, 176; film industry, 38; flexible manufacturing, 161; oil companies, 72
Spain: Basque separatist movement, 126; taxes, 140; tribalism, 126
Spanish language, 9–13, 83; in film, 27; in U.S., 13–14
Spielberg, Steven, 19, 36
Spirited Away, 35, 38
Star Wars, 24
Stead, William T., 194–96, 198
A Sunday at the Pool in Kigali (Courtemanche), 125
Super Size Me, 76
Suzuki, Ichiro, 121
Switow, Michael, 50

Tagalog, 14
Taiwan, computer manufacturers, 72
Tales from the Crypt, 28
Tarantino, Quentin, 203
taxes: in Argentina, 141; cheating on, 141; in France, 140; and globalization, 140–42; in Japan, 140–41; in Lebanon, 141; relation to prosperity, 141; in Spain, 140
Taylor, Jeffrey, 124
Teachers of English to Speakers of Other Languages (TESOL), 15
technology, and globalization, 144–205
Telemundo, 43
telenovelas, 29, 43
Televisa, 43–47
television, 39–51; cable, 43; in Cambodia,

41; cartoons, 49; in China, 43; cultural specificity in, 46–48; in Europe, 40; in Italy, 48; in Mexico, 43–47; in Nigeria, 41; program diversification, 40–42; satellite, 43; show formats, 39; UHF introduced, 40
Thailand, 198; counterfeiting, 138; tribalism, 127; sex workers, 98–99
Thompson, Kristin, 19, 21–22, 26
Time Warner, 43, 69
Titanic, 35
Tobin, Joseph, 101
Tomlinson, John, 49, 199
Toshiba, 39, 192
Toyota, 3, 68; flexible manufacturing, 159–60
trading companies, 76
transnational corporations (TNCs), 66–79; in China, 67; DaimlerChrysler, 68; definition of, 67; foreign employees of, 77–78; foreign governments and, 79; in France, 79; GM, 68; headquarters of, 79; largest nonfinancial, 68; local aspects of, 77; national origins, 77; research headquarters, 78; top twenty, 69; Toyota, 68
tribalism: in Asia, 127–28; in Austria, 127; in Brazil, 128; in Central America, 128; in Egypt, 131; in European Union, 126–27; and globalization, 121–28; gypsies, laws against, 126–27; in U.S., 128
Truffaut, François, 36
Twentieth Century Fox, 24, 36

UGPIC. *See* bar codes
Unilever, 73, 94, 201, 204
United Artists, 36
United Parcel Service (UPS), 170, 191, 191–93
United States: airlines, 71; beverage market, 73; computer services industry, 72; corruption, 132; film, 18–38; film exports, 20–35; gender roles, 100; land use, 110–11; oil companies, 72; pharmaceutical industry, 71; post office and air freight, 167–68; television, equipment and invention, 39; television programming, 40; Spanish language in, 42; tribalism, 128